HACK THE PLANET

SCIENCE'S BEST HOPE—OR WORST NIGHTMARE—FOR AVERTING CLIMATE CATASTROPHE

ELI KINTISCH

John Wiley & Sons, Inc.

Published by John Wiley & Sons, Inc., Hoboken, New Jersey
Published simultaneously in Canada

For general information about our other products and services, please contact our Customer Care Department within the United States at (800) 762-2974, outside the United States at (317) 572-3993 or fax (317) 572-4002.

Wiley also publishes its books in a variety of electronic formats. Some content that appears in print may not be available in electronic books. For more information about Wiley products, visit our web site at www.wiley.com.

Library of Congress Cataloging-in-Publication Data:

Kintisch, Eli, date.
Hack the planet: science's best hope—or worst nightmare—for averting climate catastrophe / Eli Kintisch.
p. cm.
Includes bibliographical references and index.
ISBN 978-0-470-52426-8 (cloth)
1. Climatic changes. 2. End of the world (Astronomy) I. Title.
QC903.K56 2010
551.6–dc22

2009045990

Printed in the United States of America

10 9 8 7 6 5 4 3 2 1

To my father, a righteous engineer

Contents

ON A HOT AUGUST DAY IN 2008 A TEAM OF RUSSIAN SCIENTISTS set up an experiment to block the Sun and cool Earth. The experiment was to be carried out over a 2-square-mile area of farmland near the city of Saratov on the Volga River, roughly 300 miles southeast of Moscow. Russian Federation officials provided them with a military helicopter and a truck from which engineers would release smoke for the effort.

The leader of the experiment was Yuri A. Izrael, a controversial scientist in Russia with an international reputation to match. Said to be a close confidant of Prime Minister Vladimir Putin, he was also a prominent member of the Russian Academy of Sciences. Four years earlier, Izrael had published a letter he had sent Putin, then the Russian president, in which he said global warming required "immediate action." But it wasn't cutting Russia's greenhouse gas emissions that he proposed. Instead, he suggested burning hundreds of thousands of tons of sulfur-rich aircraft fuel in the upper atmosphere, which studies suggested would lower the temperature of Earth by as much as 4°F. "We really will be able to control the climate," Izrael said at the time.

Climate scientists around the world believed Izrael's idea was at best premature and at worse dangerous. "He's a loose cannon, past his prime," said Stephen Schneider of Stanford University. After Dmitry Medvedev assumed the presidency in the spring of 2008, the government began to embrace more mainstream positions on climate science, culminating in a release of an official report the following year that omitted Izrael's proposal.

But the strong-willed climate scientist had the wherewithal to proceed with his experiment anyway. Its aim was to validate basic calculations on a small scale. Scientists set up two detectors on the ground to measure solar radiation as well as wind velocity, temperature, humidity, and pressure. At 10:50 A.M. the experiment began. The helicopter, a Soviet Mikoyan-8, began a series of passes upwind of the detectors, flying 650 feet above the ground. Following a course perpendicular to the direction of the wind, the pilot flew back and forth five times, each pass roughly 3 miles long, and the scientists released billows of smoke as they flew. Six hours later, the scientists conducted a similar experiment using smoke sprayed from the truck.

Cloudy conditions made it difficult to detect which changes in the brightness of the Sun were a result of the experiment, but close analysis of the data suggested the smoke had scattered up to 10 percent of the Sun's rays at different points in the experiment. In a paper published in a Russian meteorology journal in May 2009, Izrael and his colleagues concluded that the trial showed "how it is principally possible" to add chemical droplets to the sky "to control solar radiation." That summer, scientists conducted a more successful follow-up experiment in which they released smoke from a helicopter at an altitude of roughly 8,000 feet.

Alexey Ryaboshapko, an atmospheric chemist in Izrael's institute, said that they hoped to soon conduct even larger experiments, using airplanes, perhaps over an area roughly 10 kilometers long. "It would be a very local experiment—over Russia, only over Russia," he said. "If we are talking about implementing this geoengineering approach, the experiment must be global." Izrael aknowledged that some opponents, including colleagues in Russia, feared "negative consequences" of geoengineering. "Such fears are speculative and have no scientific basis," he said. What was needed was "an international conference" where scientists could "estimate quantitatively the degree of real or imagined danger."

It's Come to This

David Battisti had arrived in Cambridge, Massachusetts, expecting a rout, a farce, a bloodbath. So had many of the other scientists who had joined him that frigid morning from around the country. It was an invitation-only workshop on climate science in November of 2007 for which they convened at the American Academy of Arts and Sciences, an airy temple to diligence and scholarship one block from Harvard University. Battisti shuffled out of the Massachusetts morning air and into the Academy's expansive premises.

The workshop's unholy topic was geoengineering: the concept of manually tinkering with Earth's thermostat to reverse global warming. Organizers had arranged the event to find out whether respected climate scientists such as Battisti might support research into the controversial idea. In a button-down shirt opened two buttons down, Battisti poured his coffee and watched the scientists fiddle with their muffins. One couldn't take planethacking seriously, he figured, because there's no way we'll ever know enough about the atmosphere to claim we can control it. Just because the radical notion had made it from the outer fringes of Earth science all the way to Cambridge didn't mean the group was going to legitimize it, he thought.

Since the 1960s, a handful of scientists had dreamed up various schemes to intentionally alter the atmosphere on a global scale: flying enormous sunshades above Earth, creating billions of thicker clouds at sea, or spewing light-blocking sulfate pollution at high altitude to mimic the cooling effects of volcanic eruptions. Ecologists imagined brightening the planet's dark surfaces to reflect more sunlight, by spreading white plastic across certain deserts. Marine biologists explored growing algae blooms to suck billions of tons of carbon dioxide from the sky.

Each concept took a smidgen or two of sense and added scientific optimism and a dollop of whimsy. Mostly back-of-the-envelope affairs, the papers that described them included just enough observations or calculations to suggest the ideas might work. The scientists who wrote them knew the concepts were raw and with few exceptions understood them to be options reserved for worst-case scenarios. To the broader community of climate scientists, proposing even to *study* deliberately altering the atmosphere was a heretical idea.

As Battisti poured himself coffee, he saw one of the heretics standing beside the buffet table. "That guy is scary," Battisti whispered to a colleague. It was Lowell Wood, a nuclear physicist with a broad, reddish beard and a dark jacket. His wide torso was bisected by a tie featuring the periodic table of elements. From his perch at a California nuclear weapons lab, Lawrence Livermore National Laboratory, Wood had won notoriety, if not ridicule, for proposing in 1997 to control the atmosphere's thermostat by scattering chemicals in the atmosphere. He had done so in collaboration with his aging mentor Edward Teller, the father of the hydrogen bomb. Teller, whose conservative views had often put him at odds with the left-leaning scientific establishment, had advocated in the same year that geoengineering was a better way to tackle the climate crisis than the Kyoto accords.

Wood was among a handful of geoengineering enthusiasts (for lack of a better term) who had organized previous gatherings in recent years on the topic. Organized in part by Harvard University, the

2007 meeting was to bring the geoengineering true believers together with top scientists who had long dismissed the idea as a dangerous—or, moreover, a ridiculous—fantasy. "I want to get the mainstream climate community together, the brightest stars," the meeting's co-organizer, Dan Schrag, had told me. Schrag was a geochemist at Harvard who managed to know everybody in the climate community despite a reputation as a bit of an agitator. It had taken someone like Schrag, naturally, to bring together scientists like Lowell Wood and David Battisti. "I wanted to broaden the discussion," he told the scientists as they sat down in a conference room with high ceilings.

From Harvard had come scientists in geochemistry and the atmosphere, as well as a distinguished physicist wearing a small cap. MIT contributed ocean and hurricane specialists. Battisti, from the University of Washington in Seattle, was an expert on atmospheric patterns and dynamics. He told me he felt skeptical of technological solutions to massive problems such as accumulating greenhouse gases. He'd grown up with a simpler understanding of the environment, he said, regularly visiting a family dairy farm. Battisti called himself a "progressive on most issues," and had joined seventeen colleagues in petitioning the U.S. Supreme Court in a case in which they argued that the Bush administration had "mischaracterized" scientific findings they had published. You don't have to convince *me* of the severity of the climate crisis, thought Battisti. He found a chair along a set of floor-to-ceiling windows looking out on an icy patio. But if the scientists in the room called for more studies of ideas such as Wood's, it would mean endorsing a research field that had always been considered closer to science fiction.

Or, suggested Dan Schrag in his introductory remarks at the meeting, if geoengineering was only to be explored in a worst-case scenario, the decision to conduct research on it would be tantamount to acknowledging that the worst-case scenario had come or was frighteningly close. Accordingly, the slides in Shrag's PowerPoint presentation were dread-inspiring. Fossil fuel emissions were growing by

3 percent a year, he said, and China and India were only getting started burning their share of the world's coal. The level of carbon dioxide in the atmosphere seemed headed for twice the pre-industrial level, he said, and it seemed plausible that it would reach that concentration by the end of the twenty-first century. "We're not only at the business-as-usual, but we are well above all of the business-as-usual scenarios." "Business as unusual," I thought. Earth's atmosphere had warmed 1.3°F since the 1950s and was certain to gain another degree this century as the oceans warmed. The world was rallying to set up rules to regulate carbon dioxide pollution, but few in the room were optimistic that regulations passed by the United States or the international community would be aggressive enough to stem the problem.

Schrag flipped to a slide showing Antarctica. "Are the polar ice sheets vulnerable?" the caption read. "If Greenland and/or West Antarctica started to slide into the ocean, could we engineer a way to stop it?" The seasonal ice that waxed and waned on the surface of the Arctic Ocean was disappearing at an alarming rate of 3 percent a decade. "The way the Arctic ice holds on is by the skin of its teeth," said a Harvard climate scientist. Everyone in the room had heard the body of evidence and knew how damning it was. But there was a unique intensity to hearing it all at once, in a small room, with a few dozen of the world's top scientists dispensing with the niceties. The sense of desperation hung in the air like smoke from a coal-burning power plant.

Then came the would-be saviors, played by scientists, blueprints in tow. A physicist described how to use navy guns to fire droplets of sulfate pollution into the upper atmosphere, where they would reflect a small percentage of the Sun's rays, providing a modest but dependable cooling effect. By launching billions of tiny disks into orbit around the Sun, said an expert on telescopes, engineers would be able to redirect a small amount of light from striking Earth, having a similar effect. ("I got a little money from the Discovery Channel to make some of this stuff," he explained.) Modeling research had suggested that the sulfate aerosols method could be performed for a fraction of the cost of transforming the world's energy system. That technique mimicked the cooling role that volcanic eruptions played in Earth's climate. By studying previous volcanic eruptions, scientists estimated that geoengineering the upper atmosphere with this particular technique could cool Earth by as much as 4°F in a few years.

Local climates, one scientist suggested, could be "adjusted to taste." Might the aerosols method, with years of study and improvement, be a "technical pathway to Mediterranean climates" for most anyone who wanted them, as one scientist suggested? Chris Field, a prominent ecologist from the Carnegie Institution of Washington blanched slightly. (Among other problems with that particular suggestion, he said, is that wheat and other major crops require a rainy season not found in Mediterranean climates.)

Radical notions like those were why so many scientists in the mainstream have avoided geoengineering for so long. "Right now a very small number of people have worked on this for a small percentage of their time, as enthusiasts," said physicist David Keith, whose early papers on the radical concepts gave him particular authority among the armchair geoengineers. Keith was a wildly bright guy with antiestablishment leanings. He'd turned down an academic job at Princeton University to start a special energy group at the University of Calgary. There he'd made his name as an innovative energy and climate scientist, attacking more than his share of sacred cows while blessing heresies. Wind power could disrupt the weather; burning wood made climate sense—if you captured the gases you produced; and hacking the planet, though not a concept to be taken lightly, deserved attention beyond the pages of *Popular Mechanics*. Since graduate school, Keith had struggled over the question of whether studying and publicizing the idea of geoengineering would undercut efforts to reduce emissions of carbon dioxide. "A few of us are nervous to talk about this publicly," he admitted to the group.

"The engineering that dare not speak its name," mused a Harvard physicist named Bob Frosch. Sixteen years earlier, he had battled with fellow members of a federally sponsored panel who opposed his effort to include a chapter analyzing geoengineering concepts in a major national report on climate. "It was the only time things got vituperative on one of these panels," said Frosch. (The little-noticed chapter was included.) By the same token, an atmospheric scientist had told the organizers before the Harvard meeting that it should not be sponsored by the school in case the setting could be construed "as an endorsement" of the wild idea.

"This is generation zero for climate modeling for geoengineering," Ken Caldeira of the Carnegie Institution of Washington told the group when it was his turn to talk. Since 2000, the geochemist had published studies in which relatively crude computer simulations suggested that cutting the amount of sunlight received by Earth by 2 percent might counteract the warming expected in the twenty-first century. In the intervening years, he had argued for others to pursue the research while leading a small band of true believers who for years had toiled on the edges of respected science conducting geoengineering research on paper, without federal sponsorship.

This was the Geoclique, as I called them, led informally by Caldeira and Keith. Some were topflight scientists, such as Caldeira; some were knowledgeable retirees or what seemed to be hobbyists. On an online discussion group they discussed the scientific merits of various techniques and vented about the political obstacles facing their controversial field. Caldeira's expertise was the ocean, though he had been a philosophy major in college, a programmer on Wall Street, and a researcher in the rainforest. While he had gained proficiency in atmospheric science, in part because of his interest in geoengineering, his value to the nascent geoengineering cause was as much a spokesman-organizer as it was a researcher. He and Keith managed a $1.5-million fund provided annually by Bill Gates to study geoengineering.

Keith likes to think of scientists studying geoengineering as members of either the Blue Team or the Red Team, depending on their temperament and role. Blue Team members, such as Lowell Wood, have personalities that incline them to invent ways to alter the atmosphere. Keith leans blue. Russian climate scientist Yuri Izrael and his team also are solid Blue-Teamers. Red Team members, such as a plucky climate modeler named Ray Pierrehumbert, were generally skeptical of geoengineering and strove to find flaws in the blue team's work. Caldeira was bluish-purple. During his presentation he explained why he believed the sulfate technique might protect the world's coasts from the rising seas: "By dialing the radiation where you want it you can get more or less ice," he said. "If you're trying to get snow to fall on top of Greenland, this may be what you want." Having dismissed the concept of geoengineering out of hand before the meeting began, Battisti wasn't a member of either team, though his inclination seemed Blue.

At lunch, Battisti challenged Caldeira's contention that the sulfate technique would reverse the melting of the polar ice caps. "I don't know that," said Battisti, citing the model's simplistic depiction of the ocean. The best atmospheric scientists in the world, including himself, he said, simply didn't know enough about Earth's atmosphere to be making claims about how a renovation effort would turn out.

It's difficult to weigh the risks and possible benefits of planet-hacking concepts when both were uncertain. "I don't actually work on geoengineering, and I don't especially want to work on geoengineering," said Pierrehumbert. "But now that the genie is out of the bottle, I feel I have to." He shared with the group an unpublished experiment using a computer model of the atmosphere. In it, he quadrupled the amount of carbon dioxide in the sky, but kept the planet cool with a yearly dose of aerosol geoengineering. He warned that once the experiment began, a halt in the geoengineering effort—"by, say, a war or revolution"—would result in a hellish 14°F temperature jump in the tropics over three decades, bringing with it, presumably, unimaginable ecological impacts. (One climate scientist later compared the global climate addiction to alcoholism, and geoengineering to dialysis that allows the patient to continue drinking. Disrupting the geoengineering, he said, would be like unplugging the dialysis machine. So blocking the Sun's rays might buy humanity a little time, but it made cutting carbon pollution even more important, not less.)

Could scientists hope to answer the question about whether geoengineering could help to reverse the catastrophic demise of Greenland's ice sheets, if scientists found that happening? "We don't know how to model the ice sheets," Pierrehumbert told the group. "We may not have time to understand the system well enough before we act," said a Canadian postdoc.

"In the next twenty years a president may decide that he or she wants to know whether geoengineering can help prevent Greenland from melting," Schrag told me. Facing dire straits in the future, policymakers would no doubt turn to climate scientists to ask whether radical means to take control of Earth's climate could work. "Will we have done research to have a good answer or not?" Some of the scientists in the room questioned whether their field would *ever* be able to provide a sufficiently certain answer to allow society to make a truly informed decision about planethacking. Which meant there was a decent chance it could be deployed without sufficient care. "I am really darn scared," Battisti told the group. "No one wants to see this happen. No one wants to deploy this stuff."

"If we communicate to the general public that geoengineering is a tool in our back pocket in case of an emergency, we're doing them a disservice," said a Canadian policy expert. "The public will then do less to lower their carbon emissions."

Keith seemed to resent the implication. "Being silent is unethical and arrogant," he said.

Pierrehumbert looked indignant and jumped in. "There's no denying that there's a risk that this will undercut burgeoning mitigation efforts." he said. "I would ask people not to accuse others of being unethical if they are acting so as not to let the cat out of the bag."

On the morning of the second day of the meeting Battisti began to feel his resistance to studying the idea of geoengineering dissolve. That, he told me, was an alarming consequence of what things "had come to." Particularly devastating, he said, was a discussion about the low initial cost of the sulfates technique—might any one country for a few billion dollars deploy a global geoengineering program? And if that was the case, then scientists had no choice but to study it. Even if every nation signed a global ban, they felt impelled to understand the risk if rogue states took it up.

Things had come to Robert Socolow, a senior scientist from Princeton, saying that the climate problem "is a problem we are going to solve with a portfolio. If geoengineering can prove itself . . . it deserves to be in the big leagues." Things had come to former Harvard president Larry Summers, one of the most well-connected economists in the country, signaling his support for the research. Things had come to this very prominent group subconsciously moving beyond the question of *whether* scientists should start to look at the controversial idea and on to the question of *how* they would study it.

Underlying it all, said Battisti, was a sense of fear and the larger implications for the planet, for scientists, for his sense of moral responsibility. It all hit him that morning "like a horrible train wreck," he would say later. He felt himself propelled from the room out into the Academy's softly lit front vestibule, where he paced for a few minutes. On the walls hung letters written by some of the institution's most prominent members, including Martin Luther King Jr. and Albert Einstein, accepting their invitation to become members. Battisti used his cell phone to call Seattle, where it was early in the morning. His wife answered. "This meeting is scaring the daylights out of me," he told her. The choices were stark, and the scene, he said, one of eerie inevitability. "I remember having a feeling of surrealness—that the conversation didn't really happen," his wife, scientist Lynn McMurdie, says, recalling the "powerlessness" in her husband's voice. "I don't see any reason that this can be stopped," Battisti told her. Soon after he returned to the room, the scientists voted in a straw poll to support geoengineering research, with Battisti voting in favor.

And so some of mainstream climate science's leading lights had blessed geoengineering, their unholy child. Battisti felt a little numb, defeated. "It's wrong for us not to figure out a way to pursue research," he told me the next day. "But it would be incomprehensible that we deploy this." A year after the Harvard meeting, as its attendees have come to call it, he found himself in a conference room in Santa Barbara, California, with nine other scientists. He'd agreed to join a week-long exercise to map out a hypothetical ten-year research plan to understand how to hack the atmosphere with sulfate droplets. With equal parts seriousness and melodrama, the organizer of the group, a physicist named Steve Koonin, told him to imagine that "the president has just called you. There's a climate emergency." Battisti took out a pen and began to work. He'd joined the Geoclique, playing somewhere between the Red and Blue teams.

Since the Harvard meeting, almost every forum relevant to the climate crisis has reached out to embrace, if tentatively, the former pariah called geoengineering. In 2008 the British Royal Society devoted a full issue of its prestigious *Philosophical Transactions* to the topic; the following year an expert panel convened by the society called for "coordinated and collaborative" research into planethacking to augment efforts to cut carbon emissions. Its sister organization, the U.S. National Academies, sponsored a two-day workshop on the topic that same year. The Pentagon's secretive research agency, the Defense Advanced Research Projects Agency, has considered geoengineering studies. The American Meteorological Society has called for geoengineering research since, among other reasons, it could serve to "offer strategies of last resort if abrupt, catastrophic, or otherwise unacceptable climate-change impacts become unavoidable." President Obama's science adviser, John Holdren, has said that the topic is being discussed in the White House, and top officials at the Department of Energy are quiet advocates of federal spending on the concept. (President Obama's energy secretary, physicist Steven Chu, said five months into his new job that painting roofs white could have a substantial impact on Earth's climate.) Two years ago, it was possible to read the relevant literature in the field on a train from Boston to Washington. Now, publications proposing or analyzing various means of large-scale intervention appear every few weeks.

The muted volume of dissent over geoengineering research so far has been as striking as the groundswell of interest in it. The most public opposition has come in response to a handful of medium-scale efforts by scientists aboard research vessels to grow algae on the high seas. ETC group, a Canadian environmental organization, has been among the harshest critics of geoengineering, calling it uncivil "geopiracy." (In 2009 it awarded first place in its April Fool's Day "invent-a-geoengineering-scheme" contest to a plan to pull Earth away from the Sun with space shuttles.) But even ETC thinks scientists should be allowed to study the concept.

With little public opposition, into this new arena have come a variety of Red and Blue teamers alike: confident would-be geoengineers, reluctant ones, wild inventors, and senior modelers warily turning the knobs on humming supercomputers that simulate Earth's endlessly complex biosphere. Longtime Geoclique members such as Caldeira, Keith, and Wood are in demand, and out of the woodwork have come new scientists interested in the idea. The Discovery Channel filmed a one-hour segment in a series called *Project Earth* in which a scientist tried to protect ice on Greenland by wrapping it with reflective plastic blankets. A Bay Area engineer wants to float white, breathable panels on the surface of the polar ocean to reflect solar energy, and a nuclear weapons expert in Boston told me he asked the journal *Science* whether it would be interested in publishing details on his scheme to lighten the ocean's surface with trillions of tiny bubbles.

Is geoengineering a bad idea whose time has come? Driving hybrid cars, using solar, wind, and nuclear power, or storing carbon dioxide from coal plants in the ground are the conventional solutions that would reduce the amount of carbon we emit into the atmosphere. But they may or may not be enough to avert disaster. For one thing, living sustainably won't solve the problem of the carbon that has already accumulated above our heads. "Unless we can remove carbon dioxide from the atmosphere faster than nature does, we will consign Earth to a warmer future for millennia or commit ourselves to a sustained program of climate engineering," says Keith. If things get out of hand, there could be few alternatives. "The recognition that there is no other way to actually prevent further warming this century is a sobering thought and forces us to look at these options," Caldeira says.

There are two broad categories of schemes to engineer the climate. Techniques that deflect sunlight back into space before it can strike Earth's surface are the more radical and more potent variety. Mimicking the cooling effects of volcanoes and brightening clouds over the ocean are two examples that have gotten the most attention. Scientists have also envisioned launching enormous reflectors into orbit around the Sun or Earth, or genetically altering plants to make them shinier. Enhancing the planet's natural reflectivity is generally "fast, cheap, and uncertain, but it does very little to manage the carbon in the air," says Keith.

The other type of geoengineering strategies work by reducing the greenhouse effect by drawing down carbon dioxide from the atmosphere. These include growing algae in the ocean or altering the chemistry of the ocean to enhance the natural process in which it acts like a sponge to suck up carbon dioxide. "Slow and expensive, but it gets the carbon out," says Keith.

Geoengineering invites mishap by altering aspects of the climate system about which we know the least. Adding sulfates to the sky and brightening clouds rely on the role of tiny droplets known as aerosols, which have a huge but mysterious influence on climate. The carbon-sucking category of geoengineering generally depends on the global cycle that governs the planet's flows of carbon, another big unknown in various climate models. It's not even clear right now that we understand our proximity to disaster. We're not sure how ice sheets melt, or how quickly. We can't quite track the world's carbon, whether it escapes into the atmosphere from a compact car or a rotting tree stump. Over the past century, scientists have steadily realized how subtle changes in the ocean, the sky, and the continents can have profound global effects. That raises the frightening possibility of catastrophes such as droughts and stronger snowstorms or hurricanes happening with little notice or after seemingly small pushes. But, conversely, a system that is responsive to subtle perturbations raises the hope that scientists might be able to use such levers in an effort to avert one disaster or another.

Holdren, Obama's science adviser, compares the climate crisis with sitting "in a car with bad brakes driving toward a cliff in the fog." The bad brakes are the natural buffers that usually maintain the planet's temperature, which are slow to react and may be overwhelmed by the warming our pollution is causing. If we stopped our carbon dioxide–pollution binge today, at least one degree of warming would still occur, due to the long life of CO_2 in the atmosphere and the relentless warming of the oceans. The cliff is the possibility that the greenhouse gases spewing into the atmosphere will cause a catastrophe. The fog is the uncertainty that pervades climate science—the precipice could sit a hundred feet or a mile away. It clouds decisions about how severe the problem is, how much cost we should be willing to bear to avoid it, and what the repercussions might be—how steep the ravine—if we fail. Geoengineering? That's downing half a pint of Jägermeister, yanking out the car's power steering cable, and possibly hitting a tree before the cliff ever arrives, hoping the damage isn't worse than the fall would have been. Famed environmental scientist and writer James Lovelock compares the concept of geoengineering to "19th-century medicine," with all its implied ignorance.

In 2008, Colby College weather and climate historian James Fleming told me he thought climate scientists had "lost their minds" in their enthusiasm to pursue geoengineering studies. Or, as he put it later, scientists were "sincere but perhaps deluded." Indeed, humanity has never tried anything as audacious as geoengineering—unless you count our 160-year effort to take carbon from the ground and put it into our atmosphere. To cogently oppose geoengineering research, however, one has to accept one of two faulty propositions: either the problem is not that serious, or we're on our way to solving it. These days, one will be hard pressed to find many takers for either.

Which is why there's been next to no opposition as the meme has spread steadily since the Harvard meeting. Environmental groups in Washington, D.C., have kept mostly quiet on the idea, though representatives from both Greenpeace and the Natural Resources Defense Council have signaled support for regulated research. Left-wing climate blogger Joseph Romm argued in 2009 that it would be foolish to "choose an experimental combination of chemotherapy and radiation therapy that might make you sicker if your doctors told you diet and exercise—albeit serious diet and exercise—would definitely work." And yet, like ETC Group, Romm admits that "there is no reason not to do some research."

It's one thing to take climate scientists' word when they describe palpable impacts that climate change is having on the globe. It would be quite another to believe them in the future if they say they know the planet's moods well enough to reasonably predict what altering them might cause, regardless of how gently they push. Taking planethacking seriously means weighing its possible unknown risks versus the unknown risks of the planet's current, frightening trajectory. The Santa Barbara geoengineering study, which Battisti had joined after the Harvard meeting, grappled with the issue as it prepared to release its report in 2009. An early draft of the press release described reducing emissions as "the preferred Plan A" to solving the climate crisis. Geoengineering the stratosphere, it said, was "little more" than an idea that may or may not work, "a Plan B to buy time if mitigation is not succeeding." But several members of the study worried that the wording too explicitly connected the two options. The draft that was eventually published said that geoengineering "might" possibly provide planetary insurance, since "even with aggressive global efforts to reduce greenhouse gas emissions, scientists cannot rule out the possibility of rapid changes in the climate system."

In 2000, Nobel Prize–winning chemist Paul Crutzen, writing with colleague Eugene Stoermer, suggested provocatively that Earth had entered a new geologic epoch that humans had instigated. Previous epochs, such as the Eocene and the Pleistocene, were marked by natural geologic and climatological shifts such as glacial retreats or the establishment of the savannas. In contrast, they wrote, humanity's greenhouse gas problem, deforestation, the destruction of the ozone layer, and the accumulation of a variety of pollutants in the atmosphere characterized the new era. Up to half of Earth's surface has been transformed by humans. We have supercharged the rate of species extinction up to ten thousand times in the tropical rainforests. "It seems to us more than appropriate to emphasize the central role of mankind in geology and ecology by proposing to use the term 'Anthropocene' for the current geological epoch," they wrote. Barring a global catastrophe such as an epidemic or an asteroid impact, they said, "mankind will remain a major geological force for many millennia."

The advent of geoengineering takes the concept of the Anthropocene one step beyond the inadvertent impacts that humanity has already had on the climate. It could be the deliberate control of the atmosphere that will redefine our species' dominant ecological role on Earth as the Anthropocene unfolds. Perhaps there's something about us that makes it natural to pursue that course. And yet it can be unsettling to detect the hardwired urge to solidify that dominance. Even when scientists feel a moral compunction to stop what they're doing, there's a natural drive, a curiosity, an inclination to tinker that tends to override even strong ideological misgivings.

Robert Wilson, a physicist who had led the cyclotron effort at the Manhattan Project, said decades later he "cannot understand" why his strong moral misgivings did not lead him to quit the project after Germany was defeated in 1945. "Our life was directed to do one thing," he said. "We as automatons were doing it."

"When V-E Day came along, nobody slowed up one little bit," said physicist Frank Oppenheimer, brother of Robert, the head of the project. "It wasn't because we understood the significance against Japan. It was because the machinery had caught us in its trap and we were anxious to get this thing developed."

David Battisti told me he'd experienced a similar sensation of momentum upon arriving at Santa Barbara to design the world's first comprehensive geoengineering research effort. He was explicit about the comparison. "This feels like what I've read about [what] developing the bomb felt like," he told the others on the second day of the effort. "You have to do this because, God help you if you actually use it, you want to make sure it works. You hope to God this is never used but if you have to use it, you better know how it behaves."

IN 1971, THE U.S. BUREAU OF RECLAMATION COMPLETED WORK on the Kesterson Reservoir, in the western San Joaquin Valley, California. Eighty-seven miles of drains had been laid for the project, which terminated in a set of twelve man-made ponds formed with earthen dikes. The ponds were meant to receive excess agricultural runoff water. By relieving nearby farms of water with excessively high levels of nutrients, the thinking went, the plan would aid both farmers and parched ecosystems that received the runoff. "The Bureau managers argued there would be benefit to all and adverse impact to none by using the tainted but nutrient-rich drainage," wrote journalist Tom Harris.

A dozen years later, scientists explored the Kesterson Reservoir, where the drainage ended up, in a canoe. They discovered that high levels of a natural element called selenium were killing and deforming embryos and hatchlings of birds. The chemical, usually found in trace quantities, also was killing catfish, plants, and other wildlife. The selenium, it turned out, had leached out of nearby mountains. When the region's copious sunlight and dry climate evaporated water out of the wetlands, the poison became concentrated in the remaining water.

The water was removed from the Kesterson Reservoir in 1988, and after a cleanup estimated to have cost hundreds of millions of dollars, scientists found no evidence of selenium poisoning in its wildlife six years later. The U.S. Geological Survey, however, has identified fourteen other sites in the western United States where agricultural drainage with high concentrations of selenium killed wildlife.

"At best it was a cruel trick of nature," Harris wrote. "At worst it was the predictable price of arrogance, greed and tunnel bureaucratic vision."

Hedging Our Climate Bets

Think of the possible consequences of Earth's carbon binge as points along a bell curve.

In the middle, the bulging part of the curve represents the most likely outcomes: what scientists have calculated would be a warming between about 3°F and 8°F if humanity increases its greenhouse gas emissions roughly as expected. That's the broad range of responses scientists say are most likely if we increase the concentration of atmospheric carbon dioxide to twice what it was before the Industrial Revolution.

Extremes sit on either edge of the bell—what economists call the curve's "tails." Climate-change deniers, a tiny minority of scientists who reject the mainstream view on global warming, sit on the far left side of the curve. They believe that the carbon we put into the atmosphere will have minimal effects on Earth's temperature.

On the opposite side of the curve sits Marty Weitzman, an economist at Harvard University. His side is terrifying. He wonders about the most catastrophic values of "climate sensitivity," the term scientists use for this measure of responsiveness. Climate models cannot rule out a climate sensitivity of 18°F, he notes. That is, they simulate the concentration of carbon dioxide in the atmosphere rising from its current level of 387 parts per million to 560 parts per million—that's twice the preindustrial concentration and considered likely. In response, in extreme cases, the simulated planet absolutely sizzles. (It's highly "sensitive," one would say, to that dose of greenhouse gases.) Miami temperatures in Toronto, assuming both cities still exist.

To be sure, Weitzman says, the chance of an 18°F sensitivity is highly unlikely—about one in three hundred, based on a rough analysis of many climate-modeling studies. But the chances aren't zero. Or, as he puts it, the tail of the curve is "fat," since it gets to zero slowly.

We don't think about it much, but we encounter the possibility of unlikely yet devastating catastrophe every day. Will your house burn down? The chances can be plotted on a bell curve in which the height of the curve indicates the probability. When it comes to the risk of fire, most people probably figure they're in the bulging middle of the bell, where the probability of calamity is somewhere between highly unlikely and very unlikely. If one day you suffered a fire, economists would say that you ended up on the right thin edge of the curve, or the "tail." Since there is a known chance of a fire, most people buy insurance, just in case.

Weitzman asks whether we are buying the right kind of global climate insurance. The "insurance" that we buy for global warming is the price we place on carbon. (It's not a perfect analogy, but making it expensive to pollute the atmosphere with greenhouse gases, like insurance, reduces the planet's risk by creating an incentive for industries to produce energy more cleanly.) Putting a low price on carbon is like buying cheap insurance—think, low premiums—whose payouts will protect you only in case of least harmful problem, such as a garage fire that causes smoke damage. Cheap insurance won't protect you if your house burns to the ground.

Weitzman is not even sure expensive insurance will do the trick. "The probability of a disastrous collapse of planetary welfare from too much carbon dioxide is non-negligible," he writes in the stilted but devastating language of the economist. We may not be able to see the fire coming or to react in time to lower carbon emissions given all our "fuzzy" information, he says. There might be an incendiary bomb under the front porch.

What's worse, he says, is that most economic analyses of climate change exclude the chance that the climate is disastrously sensitive to carbon in their calculations of how much insurance to buy. They exclude 1-in-300 risks, neglecting the far right side of the curve. He believes, by contrast, that uncertainty about unlikely but catastrophic events is the *most* important factor to include when weighing how seriously to take climate change. Other economists, he believes, have failed to confront the capricious nature of the Anthropocene.

When it comes to the changing state of our big blue planet, scientists have a torrent of information about the middle of the curve but must get by on a drip of clues to understand the fat tails—just how bad things might ultimately get.

First the deluge. Nine months before the Harvard meeting discussed in chapter 1 had come the release of the Intergovernmental Panel on Climate Change's (IPCC) 2007 report, six years in the making. If this was another authoritative and gloomy report in a seemingly endless string, the 2,928-page document was unusually authoritative and particularly gloomy. The product of work by roughly twenty-five hundred scientists, it described the steady changes that had marked the Anthropocene's first chapter. The level of carbon dioxide in the atmosphere hadn't been this high for at least 650,000 years, having skyrocketed since 1970 and accelerating. In the 1980s and 1990s, greenhouse gas emissions rose by 1 percent per year; between 2000 and 2005 that rate had more than doubled. In 2006 humans poured more than 8 billion tons of carbon into the atmosphere.

Man-made greenhouse gases had a warming effect that the IPCC estimated at 3 watts of energy per square meter of Earth. Imagine three 1-watt Christmas lights hanging over every portion of ocean, desert, and forest and you can imagine the effect on a global system sensitive to small changes. Eleven of the previous twelve years were the planet's warmest since 1850, the report noted. The Arctic had lost more than 385,000 square miles of sea ice since 1979. Seas were rising at an average rate of 2 millimeters per year since 1961.

Ecosystems and people were already starting to feel the effects, with worrisome trends gradually becoming crises. More intense wildfires were cropping up in the western United States, with more in the offing, along with a greater risk of giant floods. Fish and birds were migrating toward the poles to find cooler temperatures, and trees were shifting their ranges. Thirty percent of coral reefs worldwide were damaged beyond recovery, with pollution, rising temperatures, and seas made sour by the carbon dioxide glut all culprits. The report equivocated in its predictions for the future, but its range of uncertainty simply spanned the unpleasantly considerable distance between bad and horrendous news. Between 120 million and 1.2 billion more people in Asia would be facing water stress by the 2020s.

Firmly in retreat by 2007, what few outspoken climate skeptics remained attacked the report as alarmist. Oklahoma senator and lead climate denier James Inhofe called a summary prepared for the public a politically motivated "corruption of science." But there was no stopping those insidious little Christmas lights, shining continuously everywhere. And as is often the case with extensive efforts covering fast-moving fields, the report was out of date the minute it was printed. It's become clear since then that the IPCC underplayed, not exaggerated, the risks of global warming. ("It is one of the strengths of the IPCC to be very conservative and cautious and not overstate any climate change risk," German researcher Stefan Rahmstorf said when the report was published.) During the 1990s, carbon dioxide emissions grew at 1.5 percent per year. From 2000 to 2007, however, the rate was twice as fast. Humans are belching out so much carbon dioxide that the emissions are, only a few years later, higher than the most pessimistic projections the IPCC released in

2007. The planet has warmed 1.4°F since preindustrial times, with at least another 1.1°F in the pipeline, even if we were to stop our carbon diet cold, completely, today.

The report delivered ominous findings on Earth's poles, where ice sheets on land hold enough freshwater to raise global sea levels by more than 200 feet. In 2009, for the first time definitively, scientists found that Antarctica was warming, by a rate of 0.1°C per decade. On Greenland, where ice sits atop 650,000 square miles of land, the rate that glaciers are lumbering toward the ocean has doubled in recent years. The 2007 IPCC had suggested that by 2100 world sea levels would rise between 7 inches and 2 feet, depending on the warming. A year later, armed with a new understanding of how glaciers move, scientists revised the range: 2½ feet to 6½ feet. And after a two-year survey of Greenland released in the middle of 2009, scientists said that Greenland's melting alone would contribute 2½ feet to the sea level rise, without even factoring Antarctica melting. The world's oceans won't rise equally as the planet's ice melts. Scientists found in 2009 that major sea level rise will include roughly 8 inches more rise along the northeastern coast of the United States than in the rest of the world. As the polar ice caps steadily disappear, gravitational effects, water densities, and ocean currents will distribute their water unevenly. Some of the largest and most well-established population centers in the world, including the eastern coast of North America, are directly in the crosshairs for the largest amount of sea level rise.

What scares scientists such as Weitzman the most are not the questions to which we do not know the answers, it's the questions we don't even know to ask. Former secretary of defense Donald H. Rumsfeld, to much derision, made the term "unknown unknowns" famous. We have only crude tools to find our unk-unk's, as the military calls them. Climate models, computer programs with a million lines of code or more, are big and clumsy, and many still feature a vast blank "slab," as experts sheepishly call it, where a living, dynamic ocean with the complex movement of currents and heat ought to be.

While models have answered some of the big questions about global warming, they use various fudge factors that stand in for a portion of the real physics representing the wild behavior of the climate system. They're only recently including rough approximations of Earth's carbon cycle, a crucial process with which the oceans, soils, and forests suck in or spit out carbon, erasing or amplifying human sins. The models also lack a realistic treatment of aerosols, the dustlike particles that fill up the atmosphere and affect clouds. And the models don't reflect the recently discovered phenomenon, yet to be rigorously confirmed, that rising temperatures at sea are probably thinning clouds, creating a feedback loop that will accelerate the warming. Current measurements and historical data can tell us a lot about the climate system, but what humanity is doing to the planet now has never been done before. We need to know what's coming, and deficient climate models make it more likely we'll be caught by surprise.

The risks of which scientists are aware—the "known unknowns"—are scary enough. One startling reality that has become clear only in the past decade is that clean air could kill us. Pollution made up of the tiny particles in our atmosphere scatters light from the Sun, cooling the planet and playing the reverse role of carbon dioxide. It was only in the first decade of the twenty-first century that scientists discovered that aerosols might counteract a third or more of the warming effect of greenhouse gases. The tiny particles—mostly sulfates from coal plants and factories, nitrates and organic aerosols from the burning of forests, and liquid fuels—contribute to the formation of clouds, which can either block sunlight or absorb radiation. University of Exeter mathematician Peter Cox calls climate change "a tug-of-war" between two pollutants: warming greenhouse gases and cooling aerosols. But scientists don't know whether the aerosols are pulling softly or vigorously. Either way, that cooling effect will diminish as we clean our air of traditional pollutants. In 2008, scientists predicted that cleaner air over South America would make droughts that happened every twenty years in the Amazon happen every other year by 2025, and nine out of every ten years by 2060.

Remember those Christmas lights that are warming the planet with 3 watts of energy per square meter from greenhouse gases? How much, in terms of energy, are other pollutants cooling it down? A recent effort by scientists with the National Oceanic and Atmospheric Administration to quantify Earth's heat budget estimated the number at 1 watt. If that's the case, then the warming we've seen is a result of the net warming effect of *two* 1-watt lights. So as the world cuts its aerosol pollution, the effect won't be so devastating.

But if aerosol pollution is masking a lot more greenhouse gas warming, it means that the global warming problem is much worse than we thought—and we are farther out on the tail of the bad-consequences bell curve than it seems. NASA climate scientist James Hansen isn't convinced by the Colorado conclusion. It might be that air pollution is blocking the equivalent of 2 watts—meaning the greenhouse effect is warming every square meter of Earth with just 1 watt. That would imply, disturbingly, that "most of the greenhouse warming is still hidden by aerosols," says Hansen. Scientists don't know whether the carbon we've dumped into the atmosphere since the Industrial Revolution could pack a lot more punch in the future as the aerosols are cleaned out of the sky.

As carbon accumulates in the atmosphere, the planet's natural cleaning system can't keep up—and may be making the problem worse. Oceans and soils suck up as much as 3.4 billion tons of carbon each year, sequestered in the molecules of wood, leaves, dissolved gases, and even held as frozen crystals. It's Earth's so-called carbon sink. Scientists had hoped that rising global temperatures and carbon concentrations would spur plant growth, increasing Earth's ability to help erase our mistakes.

But that hasn't happened. Instead, scientists fear the sink will soon begin to leak—if it isn't happening already. And the size of a carbon flood that the atmosphere may have to bear is only getting larger. Permafrost, for example, is frozen soil found in cold areas that contains stores of methane, a greenhouse gas much more potent than carbon dioxide, though it's more short-lived. Scientists recently increased the estimate for the amount of this methane locked in permafrost from 850 billion tons of carbon to 1.7 trillion tons—about two hundred times more carbon than the world emits each year. They don't know how much warming is required for the stores to release their deadly burps. "It's like a big tank [of carbon], and if you knock the valve off you'll spill a massive amount," climate scientist Christopher Field of the Carnegie Institution for Science says. "What we don't know is whether it will be a little leak or a big gushing." Initial analyses to take into account this amplifier of humanity's greenhouse gas emissions have suggested that higher temperatures will continue to erode the planet's capacity to take in carbon. The result could be up to 2°F of extra warming by the end of the twenty-first century.

The aerosols and the leaky carbon sink are clues to the central question: how much warming will the carbon we are spewing into the atmosphere actually cause? What is the "climate sensitivity" that gets Weitzman so worried? It's the closest thing scientists have to the doomsday number, expressed in the amount of heating that Earth will experience if the carbon dioxide level in the atmosphere reaches 560 ppm—twice the level that was in the atmosphere before the Industrial Revolution. The current level is 387ppm and is rising. Will getting to 560 ppm warm us another 2 degrees? Another 8? With the concentration of greenhouse gases in the atmosphere continuing to rise at the current rate of 3 ppm a year, it will only take six decades to achieve 560ppm.

Since 1870, humanity has loaded the sky with CO_2 and reached 387 ppm—and Earth has warmed 1.4°F in response. What's going to happen next? To make that calculation, scientists must grasp an incredibly complex system whose pieces include much more than the carbon we are emitting, the aerosols, and our planet's own carbon-sucking capacity. They must take into account the following processes, organized as a set of feedback loops:

- Man-made carbon warms the atmosphere through the greenhouse effect; warming that air can alter certain regional air-circulation patterns, thinning low-lying clouds (scientists think) and driving into the air more water vapor, itself a greenhouse gas.
- A warmer atmosphere means a warmer ocean, and more carbon dioxide in the air means more food for algae, which makes the ocean darker (if the plankton die and fall, the ocean becomes lighter); plus, the algae emit chemicals that may make clouds brighter, cooling the planet; hotter temperatures mean drier coastal areas, which mean more dust, which might boost the growth of plankton off the coast.
- Warmer temperatures mean drier soils, which mean more fires, which lead to darker patches of Earth that get warmer (though the smoke might cool *or* warm the planet). Warmer seas mean melting ice; when ice floating on the ocean melts, a white, light-reflecting surface is replaced with dark, light-absorbing seawater, which amplifies the warming, as does natural carbon emissions from the permafrost, rainforests, peat bogs, and tundra, caused by rising temperatures themselves.

And there are plenty of other loops to the flow diagram; the whole thing prints out on four letter-size pages—if you want to be able to read the words.

Given the miasma of factors, feedbacks, and uncertainty, it's not surprising that scientists can't agree on Earth's climate sensitivity. But they agree with Weitzman that steady, relatively modest warming throughout the twenty-first century is actually the optimistic scenario. The fat tails loom menacingly.

• • •

Since the planet might react with such ferocity, it's essential for scientists to have a plan that would allow us to limit how much we rankle it with our carbon taunting. The 1994 UN treaty under which the Kyoto protocol was passed commits its 187 signatory nations to avoid "dangerous anthropogenic interference." Predictably, there's been much debate over what these words mean. Is reducing the concentration of carbon in the atmosphere the way to go? Or is maintaining the level of Earth's fever the goal that the nations of the world should keep in mind? The European Union has an official climate target of limiting total warming since preindustrial times to 2°C (about 3.6°F)—and we are 40 percent there. Scientists writing in the journal *Nature* recently compared the target to "a speed limit on a road," noting that this amount of warming, which would be much more pronounced near the poles and on land, would mean a planet probably warmer than it had been in millions of years. Staying below 4°F may be impossible, and even if we managed to achieve it, humans would have dramatically changed the planet. According to the IPCC, a world 4°F warmer could put hundreds of millions of people under water stress in Africa and Latin America. Globally, stronger tropical cyclones would threaten lives, as would the risk of dams bursting in mountain lakes and more frequent and intense forest fires. Twenty million years ago the planet was 4°F warmer than it is now, and the sea level was more than 20 meters higher.

It's not too late for humans to make a positive difference, but the burden is incredible, especially with China and India worsening their nascent carbon habits. The failure in Copenhagen in December 2009 of the nations of the world to agree on an aggressive mandatory regime for cutting carbon emissions just underscores the difficulty of the task. Meanwhile, the proposed U.S. legislation to cut greenhouse gas emissions in the United States will likely be watered down substantially if it is going to pass the Senate at all. A decade ago, in an article about geoëngineering, David Keith threw down the gauntlet: "Humanity may inevitably grow into active planetary management, yet we would be wise to begin with a renewed commitment to reduce our interference in natural systems rather than to act by balancing one interference with another."

It will be tough enough to make the 2°C limit. What lurks in the planet's future if the temperature were to rise by, say, 7°C (roughly 13°F) in a century or two? Scientists have found no evidence that Earth has warmed that much that quickly in more than 250 million years. (The 0.8°C we've already warmed in the past 140 years is a record, too.) And so climate scientists readily admit they know very little about the possibility, though they do have a rich enough history of Earth to offer clues about what might happen. Those who spend their days uncovering Earth's violent meteorological history are among the researchers who feel the most apocalyptic about its future.

Earth is prone to violent mood swings. It can be hot, it can be cold, but the history of the past 3.5 million years shows that *little* pushes in one direction or the other have swung Earth between ice ages and periods when, as scientists like to say, "there were crocodiles on Greenland." Often the pushes are simply wobbles in the way Earth spins as it orbits the Sun. In response to the wobbles, which do not change the amount of solar energy the planet receives but rather redistribute it over the globe, a series of cascading effects are set in motion. A wobble that slightly warms the northern latitudes melts sea ice and encourages the growth of dark-colored forests, which absorb more energy. There's also carbon dioxide released as soil warms, again multiplying the warming effect. (The process can reverse when the wobble slightly cools high latitudes.) "In response to these small forcings the Earth is whipsawed through dramatic climate changes," says NASA's Jim Hansen. The "whipsaws" usually mean that Earth's average temperature can rise 18°F in ten thousand years or so.

It's tricky predicting how the 8 billion tons of carbon humans emit each year might provoke this swerving behemoth. Will the temperature rise with the smooth, steady curve predicted by the top computer models? "Sudden change and surprise are more likely," says British Earth systems scientist James Lovelock. Hansen announced in early 2008 that he felt humanity had already passed the target of acceptable risk. Comparing historical data and current measurements, he said, suggested that the risk of global climate lurches would only worsen unless humanity *lowered* the carbon dioxide concentration in the atmosphere from the current 387 parts per million to 350 ppm. "Our home planet is dangerously near a tipping point," he wrote in 2008, citing the cascade of feedback loops, each tripping the next effect in the global climate system.

The most salient and fearsome ways by which the planet could react to the carbon jolt include the disintegration of the planet's ice sheets, killer droughts, an unstoppable methane release, and the shutdown of the global ocean conveyor, which among other things delivers warm water to Europe. Each scenario has already happened, long before our industrial era, and the planet has survived. If the worst consequences of climate change caused by people play out, another geologic era will follow the Anthropocene. Humans probably won't much affect the planet anymore because there won't be many of us around.

Until very recently, climate scientists haven't paid much attention to climate disasters. "The science of abrupt climate change is relatively young," says geologist Peter Clark of Oregon State University. Politicians and the economists who advise them haven't thought much about climate disasters, either. Al Gore may touch on disaster as a possible consequence of apathy in the film *An Inconvenient Truth*, but the economists who calculate just how much climate insurance it pays for society to take out actually don't factor in worst-case scenarios very realistically. Weitzman thinks that's a big problem and the main reason why scientists had better quickly start studying geoengineering much more aggressively.

Companies are able to sell fire insurance since they know exactly how likely it is that a blaze will strike. They know that the likelihood of any one house in the United States experiencing a fire is about four out of a thousand, based on hard numbers. But difficult-to-quantify climate catastrophes stymie economic modelers.

Weitzman estimates that there's a 1 in 300 chance that the climate sensitivity is 18°F. But to obtain that number, he uses an amalgam of respected though flawed computer models and uncertain data. He admits it's a bad estimate. But there's nothing better out there. How much would you pay for insurance if you have absolutely no idea how likely it was that your house will burn down?

Other climate economists deal with this rampant uncertainty by basically factoring away the worst-case scenarios in their models. Among the most prominent climate economists is Yale University's William Nordhaus. He has calculated that the way to curb the global carbon habit is to charge emitters roughly $30 per ton of carbon, emitted, rising to $85 by 2050. That's his estimation of how much climate insurance the world ought to take out. But his calculations essentially do not factor in extreme climate change as part of their main calculation. From Weitzman's perspective, that makes those numbers potentially "unusually misleading" and "arbitrarily inaccurate."

Since the data about the worst-case scenarios are so uncertain, Nordhaus writes, economic models such as his "have limited utility in looking at the potential for catastrophic events." That is, they focus on the middle of the curves, where scientists know the most about the climate. But Nordhaus defends his model's usefulness. Worst-case scenarios, he says, only affect the conclusions he gets from his model under very specific and unlikely conditions, such as nations deciding to do nothing about climate change despite very high increases in temperature. Plus, if it turns out that humanity is unlucky and the climate sensitivity is off the charts, humanity will have a chance to "learn, and then act when we learn, and perhaps even do some geoengineering while we learn some more or get our abatement policies or low carbon technologies in place."

Weitzman acknowledges that he could do no better than Nordhaus's numbers. Without better information about a rapidly changing and capricious climate, no one can. Weitzman simply believes that Nordhaus's estimates don't accurately reflect the possibility of catastrophe—how much society ought to spend on insurance. He says that Nordhaus's contention that society could respond quickly to rising temperatures ignores the fact that the climate system includes a number of long inertias. Atmospheric carbon dioxide stays in the atmosphere for thousands of years, and once the oceans start warming and acidifying, it could take hundreds of years or more to slow these processes. "The built-in pipeline inertias are so great that if and when we detect that we are heading for unacceptable climate change, it will likely prove too late to do anything about it," Weitzman writes.

This means that understanding geoengineering as an emergency response is more important than ever, he says. Regardless of how much society decides to spend on "fire insurance" to cut carbon emissions, says Weitzman, perhaps it ought to keep around an extra large fire extinguisher. Understanding techniques to hack the planet might be a good start.

Weitzman's ideas flip the debate about climate change and uncertainty on its head. Skeptics of climate change have long used the uncertainties that plague climate science as evidence that society need not work to cut emissions more aggressively. After all, they say, the models stink and the data is murky—so the threat may not be so great. But if the uncertainty is masking a worst-case scenario, the opposite argument holds. "The less clear the science is, the greater the implied rational response to a credible threat," writes University of Texas researcher Michael Tobis. Nobel Prize–winner Thomas Schelling writes that "uncertainty regarding global warming appears to be a legitimate basis for postponing action, which is usually identified as 'costly.' But this idea is almost unique to climate change. In other areas of public policy, such as terrorism, nuclear proliferation, inflation, or vaccination, an 'insurance' principle seems to prevail: if there is a sufficient likelihood of significant damage, we take some measured anticipatory action.'"

Advocates of government regulation have long pointed to what they called the "precautionary principle" as a guide for government action in the face of uncertain information. The principle states that without scientific consensus, a potentially harmful action should not be allowed in society (such as the construction of a new chemical plant)—unless the person who wants to perform the action can prove it's safe. (It's somewhat like a codified "Better safe than sorry.") Critics say that the principle is somewhat unscientific, since it supposes that restrictions should be put on action—usually on businesses whose activities might pollute—even if there isn't hard evidence of hazard. Weitzman's focus on the fat tails—possible dangers whose size is easier to quantify than their likeliness of occurring—shows how it doesn't make sense to discount dangers just because you can't quantify them accurately. "[Weitzman]'s rigorously formalized the version of the precautionary principle," says economist Frank Ackerman of the Stockholm Environment Institute at Tufts University.

When it comes to making decisions, economists are used to having all the facts laid out in front of them. But now, facing a planet prodded by an increasingly insistent force of carbon emissions, economists like Weitzman have little choice but to make conclusions based on the *lack* of information, rather than on hard data. "All of this is naturally unsatisfying and not what economists are used to doing, but in rare situations like climate change," Weitzman says, "we may be deluding ourselves and others with misplaced concreteness."

IN 1993 SCIENTISTS STARTED AN EXPERIMENT IN A PATCH OF THE eastern Amazonian rainforest to see how drying the soil would affect the ecosystem. The idea was to simulate a dying section of forest. They built a roughly thirty-by-thirty-foot roof that prevented raindrops from dripping off leaves onto the jungle floor. Previous experiments had shown that water in soil seemed to limit the consumption of methane, a potent greenhouse gas. The scientists figured drying the soil would increase the amount of methane it sucked in from the atmosphere. The hope was that as the Amazon dried out in the future as a result of climate change, its soil might reduce the atmospheric concentration of an important greenhouse gas.

But after tallying four years of their data, the scientists found just the opposite. Soil that wasn't protected stayed moist and sucked in methane as expected. But the drier, protected soil *emitted* methane—in some cases two or three times more than the controls. In a report on the work, scientists said they could only "speculate" that termites who thrived in the dry conditions were the culprit, but they weren't sure. "We do not fully understand all of the underlying process," they wrote.

Their most salient conclusion? As climate change dries out the Amazon rainforest, there could be "unexpected biogeochemical effects."

The Point of No Return

The prospect of climate catastrophes has certainly drawn scientists' attention to geoengineering. But a focus on the worst-case scenarios is a relatively new phenomenon for them. "Tipping points, once considered too alarmist for proper scientific circles, have entered the climate change mainstream," wrote *Science* magazine's Richard Kerr, dean of U.S. climate journalists, in early 2008. He was reporting from the annual meeting of the American Geophysical Union, the closest thing to the All-Star Game for climate science. "If a very small warming makes such a difference," said Penn State glaciologist Richard Alley at a well-attended event that year, "it raises the question of what happens when more warming occurs."

Will our continual carbon dose bring a steady and predictable rise in temperature? Or will the planet's various climate systems react with jerks and cascades of runaway catastrophes? When scientists think of climate tipping points, they envision periods of gradual change followed suddenly by rapid, possibly runaway shifts. An independent report on security risks and climate change written for the Pentagon in 2004 outlined a nightmarish scenario for the future based on past climate events: temperatures fluctuated over North America and Asia, storms and floods intensified, deadly droughts across the world led to food shortages, and political unrest followed. "We've created a climate change scenario that although not the most likely, is plausible, and would challenge United States national security," the report said.

Since the report, says Peter Schwartz, one of the authors, the likelihood of abrupt changes has only increased. Each factor that would bring them about has grown worse. "Many of the significant climate factors which we thought had longer time constants are coming up faster: ecosystem movements, temperature, glacial melting, sea ice," he said. "The older perception was gradual, relatively even change in a world that was warming monotonically. Now it's clearly more extreme climate events, more often, in more places." In 2009 the UN environment program focused on tipping points in its yearly review of climate change science: "Ecosystems as diverse as the Amazon rainforest and the Arctic tundra may be approaching thresholds of dramatic change through warming and drying. Mountain glaciers are in alarming retreat and the downstream effects of reduced water supply in the driest months will have repercussions that transcend generations."

The United Nations couldn't say for certain how likely even worse scenarios are or when they might play out. Scientists don't know. Three essential truths define the climate crisis: coal is cheap to burn, carbon dioxide lasts for millennia in the atmosphere, and certainty is rare. Human activities are responsible for roughly 8 billion tons of carbon emissions to date, but it's unclear how much we'll emit in 2020. It's unclear what technologies we'll have by then, and how fast we'll grow economically. We don't quite know how much that carbon will cause warming, and it's unclear how that warming will affect the planet, whether it's a 3°F rise or much more. This chain of uncertainty binds us and restricts the ability of scientists to predict accurately whether certain disasters will happen.

One way that scientists try to cut that chain, so to speak, is by conducting in-depth, structured polls of experts. Like the flawed polls that determine the college football rankings or the Academy Awards, it's not an exact science. And like those efforts, scientists don't expect "expert elicitations," as they call them, to give hard data, just a flavor of consensus among knowledgeable people. In 2008, scientists published an expert elicitation looking at the risk of triggering various climate catastrophes—the loss of the ice sheets, the shutdown of crucial ocean currents—if Earth warmed by 4°F or more. According to an analysis of the opinions of the scientists they polled, the chance of at least one of the catastrophes occurring as a result of 4°F of warming is roughly one in six. "Are you willing to play Russian roulette with the planet?" asked Hans Joachim Schellnhuber, a coauthor on the study, at a scientific meeting in Denmark in 2009. The slide projected on the screen displayed an illustration of a man with Earth as his head, pointing a revolver at his temple.

Scientists had completed interviews for the study in 2006; since then, the outlook for these worst-case tipping points has only worsened. For example, in 2009 British climate modelers found that even as little as a 4°F rise in global temperatures could commit the planet to losing as much as 40 percent of the Amazon rainforest. Warmer temperatures kill a relatively small amount of trees initially, they found. But the model simulated that those early losses set into motion a chain reaction that steadily, and irrevocably, dried the forest. Each tree preserves moisture for its surroundings—the jungle as a whole. (Other modelers say that the British simulation was too pessimistic.)

Certainty may be rare, but one thing is clear: the more Earth warms, the greater the chance of passing tipping points. Four of them are particularly frightening for scientists: polar ice melting, widespread drought, a catastrophic methane burp, and a shutdown of key global ocean currents.

Reviewing data one evening in 1997 in his office at the University of Colorado, glaciologist Konrad Steffen came to a remarkable and terrifying conclusion about Greenland. Data from a new set of weather stations "on the Ice," as they say, suggested that summer temperatures had risen roughly 3.5°F over five years, compared to temperatures recorded over the previous three decades.

"I thought to myself, 'It's not possible,'" said the scientist, who often wears an ice-crusted beard and speaks with a deep Swiss accent. Greenland is about a third the size of the lower forty-eight United States. Its ice is two miles thick. Steffen couldn't believe that such a cold, implacable mass could warm so quickly. "I worked all night to make sure it was correct. In the morning, I had the same result." He presented it to his program officer for NASA, which sponsored the research. "The climate cannot change that fast in the Arctic," said the official. "Go back to your desk, you have an error." But Steffen soon confirmed his calculations were correct.

In the thirteen years since that evening, the situation in the Arctic has continuously worsened and scientists have come to understand how little they comprehend the changes that are transforming the ice on the world's coldest places. In the waters that circle the North Pole and Antarctica, ice floats on the surface of the ocean as so-called sea ice. One ice sheet covers Greenland; the other one is in Antarctica, where the ice is divided into two portions: the relatively stable east side of the Antarctic ice sheet, a mass with an area roughly 30 percent larger than the continental United States, and the west side of the sheet, containing about 10 percent of the ice on the continent. But the West Antarctic ice sheet is disintegrating steadily, with the potential to release enough water to raise the global sea level more than 10 feet.

The disappearing Arctic sea ice is perhaps the most dramatic symptom of a planet with a fever. It has lost nearly half its thickness since 1999, and shrank to its smallest size in three decades in 2007. As the floating ice disappears, it uncovers dark water, which absorbs more solar energy than the reflective white ice that melted. That's why the Arctic is the fastest-warming place on the planet, having warmed about 2.5°F in recent years. But as the floating ice melts, it doesn't increase the level of the ocean, just as the level of water in a glass stays constant while an ice cube melts in it. (Antarctic sea ice is not disappearing.)

Water held in ice sheets, however, does raise the level of the ocean and could therefore terrorize billions of people who live in low-lying areas. "Seven years ago we didn't think we needed to worry about ice sheets," says Peter Clark of Oregon State. That's changed quickly. The two biggest threats to world sea levels are the West Antarctic sheet and Greenland. West Antarctica sits low to begin with, but the weight of the ice sitting on top of it pushes it below sea level, so it is uniquely exposed to the ocean. Scientists call it "inherently unstable." Greenland sits higher off the water, but its southern areas sit at lower latitude then Antarctica, exposing it to warmer air temperatures.

Melting ice sheets are particularly unsettling for scientists because they have only early guesses as to why and how they're disintegrating. In 2007 scientists with the Intergovernmental Panel on Climate Change estimated a maximum rise in global sea levels as a result of the melting ice sheets of 7 to 23 inches, but that was assuming they would simply melt as the air temperature rose, like ice sculptures at a wedding. In reality, complex forces dictate their demise: the movement of water through and below them, the terrain on which they sit, ocean currents, and even wind patterns. The early stages of the worst-case scenario are already playing out, and scientists are reluctantly aware of the shortcomings of their scientific tools to understand the fast changing situation. "We can't really afford to wait ten to twenty years to have good ice sheet models to tell people, 'Well, sea level is actually going to rise to meters and not 50 centimeters,' because the consequences are very significant, and things will be pretty much locked in at that point," NASA scientist Eric Rignot says.

Right now, the West Antarctic ice sheet is contributing only 0.5 millimeter a year to sea level rise, which means it would take seven centuries for it to release its 11 feet of sea level rise on the rest of the world. But its perilous position below sea level makes particularly important the role of ice shelves, which sit on the boundary of the continent and the water. They block the glaciers from the ocean "like flying buttresses on a cathedral," says Peter Clark.

When the Larsen B ice shelf collapsed off West Antarctica in 2002, the nightmare of the whole thing one day gradually sliding into the ocean piece by piece became much more real. After the Rhode Island–size shelf broke up, the glaciers behind it began moving eight times as quickly as before. Subsequently, five smaller shelves on the peninsula also have gone to pieces. New York University physical oceanographer David Holland is particularly frightened about how warm subsurface ocean currents may be attacking the ice shelves. In some of his computer simulations, warm water strikes the Ross ice shelf, the continent's largest, suggesting a new threat to the ice.

What scientists have seen on the edges of Greenland has also terrified them. The mighty Jakobshavn glacier on the island's southwestern coast, the fastest-moving conveyor belt for ice escape, doubled the pace of its march toward the sea between 1997 and 2003, according to NASA. In 2008 Holland discovered fisheries data that showed that a shifting surge of warmer water toward the glacier, due to shifting Atlantic Ocean winds, occurred just as the accelerated melt happened. The shift may not be directly related to global warming, but the rising global temperatures of the sea can only exacerbate the problem. "That warmth now has an obvious way to get at the ice," Richard Kerr wrote in *Science* magazine in 2008.

Could scientists focus geoengineering efforts locally to rescue the poles if the worsening situation accelerated? Most of the early studies on geoengineering ask very general questions—how a particular technique may or may not limit the rise in temperature, or how its effects may work around the world. Climate tipping points are triggered by rising temperatures, so the main way that humanity might avoid triggering them is to avoid warming the world too much.

But as planethacking research has slowly matured, scientists are increasingly tackling what Ken Caldeira calls "the screwdriver problem." Caldeira met in the spring of 2009 with officials from the Defense Advanced Research Projects Agency at the Pentagon. They advised him that their strategy in conducting research was to avoid "designing screwdrivers" and instead to figure out what kind of screw you want to turn and then come up with a tool to do so.

If the screw is cooling the poles and saving the ice sheets, then the focus would be providing regional cooling targeted to one or both of the poles. Putting sulfur pollution or other particles that scatter sunlight over the Arctic may or may not be an effective way to cool it. Michael MacCracken of the Climate Institute in Washington, D.C., believes that injecting such particles into the upper atmosphere or the lower atmosphere might make a difference, perhaps most effectively by injecting "only during the summer months," as he wrote in 2009. He imagined an effort "fine tuned" to respond to the changing angle of the sun's rays, fluctuating patterns of sea ice and levels of pollution in the sky. "Holding off on geoengineering until all is nearly lost is like waiting to help those facing severe climate impacts until they are malnourished and near death," he wrote, advocating "aggressive research and even a low-level start to geoengineering" to stave off an emergency situation. "Of course, we may only know the situation is an emergency one well after that point is reached," he added.

Others disagree that the Arctic could be locally cooled. It gets much less sunlight than the tropics, so blocking those rays there would have less effect there than elsewhere. Also, the atmosphere gets mixed up near the poles more intensely than elsewhere. So geoengineering with sulfate droplets as a kind of cap above the Arctic may not work. The particles might get washed out more quickly than those added farther south. "You can't put a yarmulke on the planet," says Alan Robock, an atmospheric scientist at Rutgers University.

Most of the heat that arrives in the Arctic comes via the atmosphere and the ocean, delivered from the tropics. That's why Alan Gadian of the University of Leeds in the United Kingdom suggests cooling the tropics to protect the poles. "It's like there's a metal bar extending from the tropics to the Arctic. Then if you cool the hot part of the bar, the source of the heat, you'll send less energy to the Arctic." His method of choice for lowering global temperature is to whiten tropical clouds. MacCracken feels cloud-whitening could even be used to cool the warm ocean currents near the poles that scientists like Holland believe to be threats to the ice sheets. (An even more innovative idea of his is to *destroy* sea ice using special ships called icebreakers near the ice sheets to reduce the floating ice's "insulating effect" on the warm currents below that threaten ice sheets.)

Scientists also have proposed other radical fixes. Glacier expert Doug MacAyeal, from the University of Chicago, mused in a short paper in 1984 that scientists could drill holes in the ice shelves and pump seawater from below on top of them, weighing them down so that they would lodge against the seafloor and stop moving. Another way to protect Greenland's ice might be to directly cool the waters of the Arctic by making them more reflective. After seeing *An Inconvenient Truth*, starring Al Gore, at a theater near her home in Palo Alto, California, an engineer named Leslie Field decided to put her successful expertise as a Silicon Valley entrepreneur to the task of protecting the floating sea ice. She attended seminars on the climate problem at nearby Stanford and met with a variety of professors to understand the problem. Previously, some had proposed white plastic floating barriers. But the problem with those barriers is that they trap heat beneath them, preventing the water from cooling. So after months of trying different materials, Leslie came up with silicon beads—"like grains of sand," she says—and other similar materials that have within them spaces of air or other inert gases. In tests she has conducted in California's Sierra Nevada mountain range, she has found that ice protected by her material, held loosely in cotton sacks, loses nearly three times less mass when exposed to sunlight as unprotected snow or ice does. Now she's looking for money to support her nonprofit, known as Ice911, as well as collaborators among ecologists to find out how her material might affect Arctic ecosystems. "I've gone as far as you can go on a shoestring," she says.

Roughly fifty-two thousand people lost their lives as a result of heat stress during the 2003 heat wave in Europe, the deadliest climate-related disaster of the modern era. And yet its consequences affected even more people by harming crop yields. Corn yields in Italy and France each dropped by roughly a third. French wheat and fruit harvests declined by 21 percent and 25 percent, respectively.

The outlook for farmers—and the rest of us, who rely on their labors every day—is unpromising as we race into our greenhouse century. For a number of years, scientists were optimistic that rising temperatures and more carbon dioxide—the main food that plants need to grow—would help plants and therefore boost agriculture. However, while that may be the case in some regions, recent findings suggest that the greenhouse world of the twenty-first century will mean drier soils in already arid areas, less rain in places that need it, well-fed weeds, and more persistent pests.

The first problem is that scientists believe the hotter world will be drier in dry areas and wetter in wet areas—which means more severe droughts and rougher storms. And the changes could come quickly. To a geologist, the Dust Bowl, the severe drought that rocked the U.S. Midwest in the 1930s, is an "abrupt" event; to a farmer, it's a cautionary tale. But four megadroughts that rocked the planet between the years 900 and 1600 were even worse, lasting twenty-three to forty-one years each. They decimated the U.S. West and plunged more than 60 percent of it into drought. One, which lasted from 1270 to 1297, led to the abandonment of the Anasazi cliff dwellings of the U.S. Southwest.

Such disasters happened long before humanity began its carbon binge, so scientists are unsure how a warmer, more carbonaceous sky could affect the chances of megadroughts in the future. Models and data from the past are sketchy, but the megadroughts seem to have been caused by relatively cool temperatures in the Pacific, a phenomenon called La Niña. A federal report published in 2008 concluded that it was

> disquieting to consider the possibility that drought-inducing La Niña–like conditions may become more frequent and persistent in the future as greenhouse warming increases.

> We have no firm evidence that this is happening now, even with the serious drought that has gripped the West since about 1998. Yet, a large number of climate models suggest that future subtropical drying is a virtual certainty as the world warms and, if they are correct, indicate that it may have already begun.

To make matters worse, even if rain patterns manage to remain favorable for agriculture, the rising temperatures could be just as deadly. Breeders around the world have developed crops that perform best for relatively narrow ranges of temperature, and gradually rising temperatures could threaten the ability of billions of people to get food easily. "In temperate regions, the hottest seasons on record will represent the future norm in many locations," wrote David Battisti and Rosamond Naylor in *Science* in 2009. After analyzing twenty-three climate models, they concluded that there were two dismal possibilities. Either today's hottest temperatures will become the average temperatures tomorrow, or, as Battisti and Naylor wrote, temperatures will be "out of bounds hot": even the coolest seasons will be hotter than the warmest are now. It's a crucial issue for farmers because for every 2°F increase in seasonal temperature, crop experts say, most major grains lose 2.5 percent to 16 percent of their yield. It's not clear that crop breeders will be able to keep up with the rising temperatures, providing plants that can withstand the hotter, drier conditions in time. This is especially true in the poorest nations.

Even with enough rain, there's already evidence that temperature spikes can be deadly for farmers. In the Sahel, the semiarid region that stretches from Senegal to Eritrea across central Africa, a three-decade-long climate disaster from the late 1960s to the 1990s killed millions of people as livestock and crop yields plummeted in the severe drought. In the past fifteen years, wrote Battisti and Naylor, rain has returned, bringing relief. But, ominously, the temperatures during growing season have trended upward since 1980, contributing to low crop yields for corn, millet, and sorghum.

For Battisti, a possible devastation of agriculture as a result of the warming globe is a central reason he's become interested in geoengineering, since controlling the severity of temperature changes in the coming decades will be crucial for protecting farmers.

Beneath the ocean lies one of most vexing dangers that our warming planet faces: methane, the noxious gas found in natural gas and flatulence. In land, locked in permafrost, there are roughly 400 billion tons of it. Even more can be found in the oceans, held as mysterious icelike structures known as hydrates. Methane lasts roughly only nine to fifteen years in the atmosphere, but averaged over a century its greenhouse warming potential is twenty times higher than CO_2. The release of underwater methane would be the mother of all feedback loops, multiplying a frightening warming—say, 11°F over the next fifty years—into a hellish one of 32°F.

It's probably happened before. Fifty-five million years ago, some scientists believe, crystals of methane frozen close to the ocean floor began to suddenly melt as the ocean warmed gradually, releasing trillions of tons of methane over centuries. As the methane supercharged the greenhouse effect, the planet warmed, releasing more methane in a deadly feedback loop that caused the planet's temperature to skyrocket by 18°F in centuries; scientists can't say for sure how quickly. The overheating caused an estimated two thirds of the species in the ocean to become extinct. Ocean circulation patterns shifted, and trees rapidly changed their ranges. Amid the cataclysm, the warming may have jump-started the evolution of a variety of mammals, even the species of primates that eventually evolved into humans. "Life on Earth was transformed almost as much as by the asteroid hit 10 million years before that wiped out the dinosaurs," wrote journalist Fred Pearce in *With Speed and Violence*, his 2007 book on climate tipping points.

Scientists don't know how warm it has to get on Earth to unleash the methane beast. Since they are not sure how much of it there is, it's hard to estimate how serious a threat it poses. The methane stuck in the permafrost may well be leaking slowly right now, already amplifying the warming our planet is experiencing, but scientists haven't come up with any nightmare scenarios for its abrupt release. The gas locked beneath the ocean, however, may well be released by landslides on the floor of the sea, changes in the sediments in which it sits, or subtle shifts in temperature and pressure that might cause the icelike methane hydrates to bubble out of the ocean. (Gas companies look for such releases so they can capture the methane, the main component in natural gas.) But every degree we warm the planet means we are pushing our luck that much further.

In the late 1980s geoscientist Wallace Broecker drew one of the seminal pictures in modern climate science: the so-called Great Ocean Conveyor Belt, which moves water between the Atlantic and Pacific oceans at a rate faster than the flow of the Amazon River. Broecker posited that the conveyor brings warm water from the tropical Pacific through Southeast Asia and around the bottom of Africa up into the Atlantic. In the North Atlantic it provides its warmth to Great Britain and northern Europe. Then, as it cools, it falls as much as 1,300 feet and travels south to Antarctica and below Australia until it rises near Alaska to complete the conveyor. Dozens of key eddies, currents, and whirlwinds that dictate various local climates are left out of the picture, but the concept accurately portrays a basic truth: by delivering tropical warmth to the North Atlantic, the ocean's conveyor makes the United Kingdom habitable despite sitting 12 degrees latitude closer to the North Pole than Boston. As scientists learn how delicate the system is, they become concerned that whacking it with greenhouse warming might shut it down.

When Broecker suggested how such a conveyor operated, he said that it helped explain how 12,900 years ago, just as the last gasp of the glacial period was ending, the world had experienced an almost instant cold snap. His hypothesis, also proposed by geophysicist Robert Johnson, explained how a river of fresh meltwater surging down the Mississippi valley toward the Gulf of Mexico abruptly halted. After that, scientists later discovered, an enormous glacial lake of freshwater sitting in what is now southern Canada suddenly burst through toward the east, eventually dumping out in a cataclysmic flood into the North Atlantic. That massive surge of freshwater into the ocean was like shoving a broomstick into the motor that drives the global conveyor at just the right point.

It shut the whole thing down. Temperatures abruptly plummeted by 3°F to 5°F in the tropics and by 54°F near the North Pole. It was a historic cold snap—almost geologically comic. Rivers in South Africa shifted direction wildly in the chaos, as did trade winds in the tropics. All the while, the planet grew suddenly dustier and much drier as the ice spread from the poles toward the equator. Thirteen hundred years later, in another bizarre shift, temperatures went back to their previous levels. Penn State glaciologist Richard Alley believes that might have happened in a single year—"perhaps even a single season." Since overall Earth was on a warming trend, what had caused the freezes? Frighteningly, scientists say it was simply the nature of chaotic systems undergoing change—like riots that seem to crop up at random during a revolution. The planet, Pearce wrote, was like a "drunk [on] a rampage."

With cinematic license one would expect, the film *The Day after Tomorrow* depicted a world almost instantly frozen after a shutdown of the conveyor. Its protagonists trudged their way through a New Jersey covered completely in yards of ice and buffeted with massive snowstorms. In real life, a panel of scientists in 2008 put the odds at one in ten that it could happen in the twenty-first century. For now, if the conveyor were to slow in the future, which scientists believe is possible, it could have a slight cooling effect on Europe, offsetting the warming the Continent would experience otherwise due to rising global temperatures. But scientists are concerned enough about the issue that in 2004 the United Kingdom and the United States began a monitoring program of key sites, including the North Atlantic, the Bahamas, and the waters off the western coast of Africa.

Current computer models say it could take one to two centuries for the shutdown to occur, and that the warming required to provide enough freshwater from glaciers into the Atlantic would require a carbon dioxide concentration in the atmosphere roughly three times higher than today's level. "Of all the things we have to worry about, it's down the list," says Broecker. "It's reassuring that we find it to be unlikely, but it's not reassuring that we don't have much confidence in the models that say that," said Jean Lynch-Stieglitz, a paleoceanographer at Georgia Institute of Technology. How might scientists restart the conveyor if it were to begin to slow down? "You really don't know what to do," she says. "Anything you can do to keep the climate the way it has been over the last eight thousand years would be your best bet."

When scientists have only a vague sense of the exact risks of various calamities, says geochemist Dan Schrag of Harvard, it's cause for more concern, not less. I asked him in the spring of 2009 about "high-impact–low-probability" events such as the possible disintegration of the West Antarctic ice sheet. "That's not the right way to think about it," Schrag admonished me. "They're high-impact–unknown-probability events." It's only recently that scientists have started thinking seriously about worst-case scenarios, he said, and the research efforts that exist are small and poorly coordinated between disciplines. "The oceanographers are not talking to the ice guys." This is all to say, the chance of tipping points could be much greater—or their onset, much sooner—than we estimate, which, he said, makes even greater the imperative to study geoengineering approaches and be better prepared in case we need them.

ON SEPTEMBER 26, 1991, EIGHT CREW MEMBERS ENTERED Biosphere 2, a glass-enclosed greenhouse covering more than three acres that had cost $200 million to build. Designed to allow study of ecosystems and human habitation, the sealed facility included five encapsulated ecosystems, including a rainforest, savanna, grassland, and a million-gallon ocean. A team of scientific advisers from around the world had helped design each one, and the giant structure, situated in the desert outside Tucson, Arizona, was sealed in with four thousand species of plants and animals. Known as Biospherians, the eight-person crew was to be the first to inhabit the structure on a series of missions that were to last a hundred years.

Problems started almost from the beginning. The most serious was an unexpected and mysterious loss of oxygen from the air over sixteen months, from a concentration of 19 percent down to 14.5 percent. The crew suffered headaches and fatigue. They were put on medication before oxygen was injected into the system, which was supposed to remain closed for the duration of the exercise. Eventually scientists realized that the rich soil in which the crew had grown their plants was consuming the oxygen, while the carbon dioxide it was producing was undergoing an unexpected chemical reaction with the concrete walls. "The would-be Eden became a nightmare, its atmosphere gone sour, its sea acidic, its crops failing, and many of its species dying off. Among the survivors are crazy ants, millions of them. . . . The crew lost weight, got sick and began to grow paranoid about food theft," wrote the *New York Times* in the aftermath of the famous experiment. A number of scientific papers have been published based on the experiment, and some say that the experience could prove useful for understanding the challenges of building terrariums in space.

"It's such a perfect example of human arrogance," ecologist Penny Chisholm said. "Thinking we can just set up an artificial ecosystem and have it go along in perfect long-term equilibrium."

The Pinatubo Option

Most scientific papers appear with little fanfare and dissolve into the archives. A paper Paul Crutzen wrote in 2005 caused a furor a full year before it ever appeared. The document stated that blocking the Sun to cool Earth could be relatively easy and cheap. That's what scared atmospheric scientist Meinrat Andreae so much, and that's why he led an effort to convince the Nobel Prize winner not to publish the paper.

Crutzen is a Dutch chemist, a stubborn one with a knack for unorthodox ideas and for making trouble. Andreae was his colleague at the Max Planck Institute for Chemistry in Mainz, Germany. In e-mails sent during the autumn of 2005 to various colleagues, Crutzen had distributed a paper calling for research into the controversial method, which would involve polluting the upper atmosphere to increase the planet's brightness. His paper argued dispassionately that 5.3 million tons of sulfur pollution per year, delivered to the stratosphere by plane or other means, could compensate for the twenty-first-century warming that was coming. The cost? Less than $50 billion per year. Crutzen knew that even just talking about the idea would be controversial. "I sort of knew that hell would break loose," he told me later.

Halting our carbon binge was "the preferred" option to attempting "climate control," wrote Crutzen. "However, so far, attempts in that direction have been grossly unsuccessful," he wrote. Hopes of change in this regard he called "a pious wish." Crutzen mentioned that efforts to clean the lower atmosphere of traditional pollutants would soon accelerate global warming by removing the so-called pollution mask. Crutzen essentially proposed swapping a deliberate man-made cooling haze for an inadvertent one. The haze would consist of microscopic droplets of sulfuric acid, formed after sulfur dioxide gas was released in the atmosphere. There was a natural analogue: Mount Pinatubo in the Philippines, which had spewed out the equivalent of 10 million tons of sulfur in 1991 as sulfur dioxide, cooling Earth by half a degree Celsius the following year. Geochemist Ken Caldeira would soon dub Crutzen's proposal the Pinatubo Option.

Andreae told Crutzen that he should not publish the paper. Crutzen made sure to mention some of the side effects of the technique and noted that cutting emissions was the main priority. But Andreae thought such caveats were insufficient. There also were subtle differences in the way the two men viewed humanity's place in the environment, said Carl Brenninkmeijer, a mutual friend. "Andreae, he's German; he loves birds, the outdoors. Crutzen's from Amsterdam; he's much more pragmatic. There is water rising—he is someone who builds the dyke." Andreae was an early Red Teamer; Crutzen, among the most prominent of the Blues.

The Pinatubo Option wasn't a new idea, but it had never been championed by a scientist of Crutzen's stature. Crutzen's Nobel Prize–winning research on the ozone layer had made him practically a hero of environmentalists. With that kind of credibility, said Andreae, the potential publication of the paper would be "irresponsible and immoral." ("If he says it, then it's 'Nobel Prize Winner Has Solution to Climate Change: Put Aerosols in the Stratosphere,'" Andreae told me later. "If you or I said it, it's 'Crank Scientist X Has a Kooky Idea.'") It was the kind of thinking that had gotten humanity into the global warming problem in the first place, continued

Andreae, with dire side effects in the future. "Geoengineering is like a heroin addict finding a new way of cheating his children out of money."

A dozen or so famous scientists argued the issue in a lengthy e-mail exchange. Crutzen's former student Mark Lawrence, a scientist at Max Planck, convinced his mentor to add more discussion of side effects to the paper. Andreae was still opposed. "Would you like giving George Bush control of the world's climate?" he asked. "These are desperate times," Crutzen responded, and submitted the paper to the influential journal *Climatic Change*.

The expert reviewers said, oddly, that the paper was scientifically solid but inappropriate for publication. "It was a weird review," said the journal's editor, Stephen Schneider, who leans Blue. He sent it to Ralph Cicerone, the president of the National Academy of Sciences and a friend of Crutzen's, to resolve the dilemma. After much negotiation, it was agreed that the paper would get published as long as it would appear along with essays offering various perspectives on the controversial idea. "The Geoengineering Dilemma: To Speak or Not to Speak" was the title of one. It included the following text: "Geoengineering is being discussed intensely, at least outside of the formal scientific literature, and it is not going to go away by ignoring it or refusing to discuss it scientifically."

News of the paper's publication appeared on the front of the science section of the *New York Times*. "The source of the proposal was almost as remarkable as the idea itself," wrote a reporter in *Science*. "When I saw the paper I thought of 'A Modest Proposal' by Swift," said David Battisti. "Some kind of joke." Crutzen's instincts in the past had earned him a reputation as a trailblazer with famous scientific papers on nuclear winter, the effects of burning rainforests, and the role of droplets in the atmosphere. This time his impulse launched a powerful meme. More than any other event, the publication of his paper brought the idea of geoengineering into the scientific mainstream. Its appearance eventually led to a dozen or more geoengineering papers in the following three years, the Harvard conference that Battisti had attended, a geoengineering meeting at the National Academy of Sciences, the secretive Santa Barbara confab, and the Royal Society report on planethacking. "It's hard for me to imagine that the issue would have exploded without Paul's statement," Lawrence said. Cicerone said the paper "had much more of an effect than I expected."

The Pinatubo Option is a bad option for mitigating global warming the way that war is a bad option for resolving global conflict. It is a tourniquet; its application involves accepting great risk in order to preserve life. Diehard Red Teamers say manipulating our mysterious skies is too arrogant and distasteful even to consider. But we have no choice but to examine it, and closely. To ignore what scientists believe to be the most effective method we have for cooling the planet fast might mean foreclosing on an option our society might one day need.

We wouldn't need any outlandish technology to try it—just some jet aircraft, naval guns, balloons, or aerosol tanks to get the gunk to the stratosphere. Reputable studies have suggested that implementing the approach would cost mere billions per year, initially, to offset the warming caused by all current CO_2 emissions. It could compensate for the warming caused by skyrocketing carbon pollution in months. The technique could be stopped on short notice. But the fact that the option could make the sky bluer or whiter, depending on what was to be added to the sky, only hints at the hubris involved. Its use could cause a number of frightening environmental side effects, not to mention the geopolitical implications. And yet there is an inescapable and disturbing possibility that the impacts of global warming without geoengineering could be worse.

Making a man-made volcano is surprisingly feasible. To mimic the effect of volcanic eruptions, engineers would produce sulfur dioxide, a common chemical ingredient, and loft it into the stratosphere, where normal chemical processes would convert the gas into droplets of sulfuric acid—an aerosol. Spread out across the globe and too high to be washed out by weather, the sulfuric acid haze would provide its cooling effect on a global scale for as long as two years. Modeling studies and data from volcanoes suggest that the Pinatubo Option could lower global temperatures by 5°F or more. (A tenth of an ounce of sulfur in the high atmosphere roughly offsets one ton of carbon, scientists have calculated.) Deployed very aggressively, the option would become apparent almost instantly, as droplets would cause a perceptible bleaching of the sky. The whitening effect happens because the droplets would scatter all visible wavelengths of light equally, making a whitish color. (The natural sky looks blue because air happens to scatter the Sun's rays at the wavelength that produces that color.)

Since Soviet climate scientist Mikhail Budyko first proposed the Pinatubo Option in the 1970s, scientists have dreamed up a variety of other ways to block the Sun's rays. In its 2009 report on planet hacking, the Royal Society noted that these schemes, listed with their dates and inventors, have included, as it described them:

- a refractor made on the Moon of a hundred million tons of lunar glass (Early, 1989);
- a superfine mesh of aluminum threads, about one millionth of a millimetre thick (Teller et al., 1997);
- a swarm of trillions of thin metallic reflecting disks, each about 50 centimeters in diameter, fabricated in space from near-Earth asteroids (McInnes, 2002);
- a swarm of around ten trillion extremely thin high-specification refracting disks, each about 60 cm in diameter, fabricated on Earth and launched into space in stacks of a million, one stack every minute for about 30 years (Angel, 2006).

Scientists have come up with schemes to reflect light away from Earth in four different places. Space schemes would do it above the atmosphere. The Pinatubo Option works in the upper atmosphere. Brightening clouds would reflect light from the lower atmosphere. On Earth's surface, making plants shinier, painting roofs white, lightening the ground, or whitening the ocean would make the planet reflect more solar energy to space.

It might seem obvious that preventing sunlight from striking Earth would cool the planet, but scientists were skeptical that it would work at all when astrophysicist and enthusiastic Blue Teamer Lowell Wood discussed the Pinatubo Option at a meeting in Aspen in 1998. (An acolyte of Edward Teller, Wood played to type, goading on the audience, largely scientists and leftist environmentalists, by titling his talk "Geoengineering and Nuclear Fission as Responses to Global Warming.")

Atmospheric scientist Ken Caldeira, sitting in the audience, was unconvinced that Wood's plan could be feasible. While carbon dioxide warms Earth evenly over its entire surface, he thought, blocking sunlight would cool it in patches. Carbon dioxide warms the planet twenty-four hours a day, for example, while the Sun heats the planet only during daylight hours. Caldeira and his coauthor, Livermore lab scientist Bala Govindasamy, suspected that the technique might cause nighttime and daytime temperatures to be more similar and the seasonal differences to be less pronounced, wreaking possible ecological havoc. The most serious reason why the system may not work was that turning the temperature down on the whole planet might not necessarily cool the Arctic, Caldeira thought; the poles receive sunlight only half of the year. He figured that the effect of the technique might lead to instabilities even worse than those accompanying global warming.

Caldeira and his coauthor did the first-ever computer modeling of the controversial idea. They used a relatively simplistic model that simulated the atmospheric cooling effects of the stratospheric gunk by simply turning down the Sun's intensity by 2 percent. A simulation of Earth with doubled CO_2 and the geoengineering at the same time revealed, to their surprise, that the geoengineering very effectively compensated for the warming that the carbon dioxide caused. According to the model, the sea ice sitting on the North Pole was preserved; the "melting of Greenland and Antarctic ice caps and the consequent sea level rise is less likely to occur in the geoengineered world," they concluded in a 2000 paper. The main reason, it seemed, was that since the poles were the fastest-warming parts of the planet, the reduced cooling they received from the Pinatubo Option had a greater effect there. Maybe Wood had been right—or lucky.

One Wednesday evening, April 5, 1815, British lieutenant governor Sir Stamford Raffles, stationed on the island of Java, heard what sounded like a series of cannon blasts in the distance. Concerned that one of his posts was under attack, he dispatched troops to march toward the noise. The rumblings turned out to be the prelude to the largest volcanic eruption in recorded history, the explosion of Mount Tambora on the island of Sumbawa, now part of Indonesia. The rumbling continued for five days, and then ash began to fall. The rajah of Sanggir told Lieutenant Owen Philips that he had seen "three distinct columns of flame burst forth . . . and after ascending separately to a very great height, their tops united in the air in a troubled and confused manner . . . the whole mountain next [to] Sang'ir appeared like a body of liquid fire." A violent whirlwind followed, destroying nearly every house in the village. Explosions were heard as far as 1,600 miles away, pillars of fire soared miles into the sky, and a mountain 2½ miles tall turned into a smoldering river of lava that caused the death of more than 71,000 people by fire, disease, and starvation. A 12-foot tsunami hit nearby Indonesian islands, and ash fell for three days over an area of 170,000 square miles.

On the other side of the globe, however, the eruption affected the lives of many millions more. The volcano had spurted an endless stream of sulfur dioxide gas 21 miles high, which formed 200 million tons of sulfuric acid droplets in the atmosphere. Spectacular sunsets were the first effects that the droplets caused. Observers in London marveled at streaked glows of orange, red, purple, and pink, with some clouds reflecting colored light half an hour after sunset. Astronomers of the period said that only the brightest stars were visible in the night sky. The Sun looked so dim that passersby could look directly at it, and even make out sunspots; the haze of droplets were known by Westerners as the "dry fog." Trapped in the stratosphere, neither winds nor rain could wash it out.

Eighteen sixteen was to be known as "the year without a summer." Scientists say the average temperature dropped in the Northern Hemisphere, for example, roughly 1°F—the coldest known year since comparable weather records were begun in 1750. Snow fell in June in Maine and upstate New York, and the length of the growing season in the northeastern United States plummeted from 130 days to about 70 days. "On the 10th of June, my wife brought some clothes that had been spread on the ground the night before, which were frozen stiff as in winter," wrote farmer Chauncey Jerome from Plymouth, Connecticut. Corn, cucumber, and hay harvests were decimated across the world, worsening famine, rioting, and disease in the aftermath of the Napoleonic Wars. The global melancholy is said to have inspired Mary Shelley to write *Frankenstein*. "The bright sun was extinguish'd," wrote Lord Byron from Geneva that year in "Darkness," a poem that captured Europe's grim mood. "Morn came and went—and came, and brought no day/And the men forgot their passions in the dread/Of this their desolation."

Before there was Tambora there was Mount Etna, in 44 B.C., which Plutarch suggested was responsible for famine in Rome and Egypt. Ben Franklin linked the 1873 eruption of Lakagigar, in Iceland, to an abnormally cold subsequent summer and winter. Pinatubo, which erupted in 1991, has become a poster child for the sulfate method because the moderate eruption was carefully monitored. It spurted only a tenth as much rock and material into the atmosphere as Tambora and only a sixth as much sulfur.

Mimicking the cooling effects of a volcano's gunk might work with any variety of methods. A unique company in Bellevue, Washington, has proposed what it calls a stratoshield, or "hose to the sky." Intellectual Ventures, run by former Microsoft executive Nathan Myhrvold, believes that three 18-mile-long hoses supported by blimps could form a delivery system for the stratoshield, which would pour a hundred thousand tons of sulfur dioxide each year into the polar atmosphere. Its job: "rescue the Arctic ice cap and tundra from catastrophic warming" by blocking a tenth of incoming solar radiation north of 60° latitude. ("Fixing Global Warming with a Helium Balloon and a Couple of Miles of Garden Hose" read an approving headline on the *Wall Street Journal* opinion page.)

Others have proposed more complex methods of delivery. Rutgers University professor and enthusiastic Red Teamer Alan Robock, an expert on volcanoes, has suggested that supertanker planes could deliver the chemicals to the sky, where jets would distribute them in the stratosphere. (Many planes can't reach the stratosphere, so fighter jets may be the best bet, requiring 180,000 flights per year at an annual cost of more than $4 billion.) He also envisioned building a 70-mile high tower using carbon epoxy composite materials. Balloons might be used to get the gases into position. Robock has calculated that nine million weather balloons full of sulfur pollution, each about 15 yards wide, would be required every year to roughly compensate for a significant warming. One side effect would be what a federal report called "an annoying form of trash rain"—220 million pounds of collapsed plastic balloons falling out of the sky each year after they deliver their payload.

Might scientists commandeer Earth's most prodigious—if unreliable and destructive—sulfur producers? "Has anyone considered management or engineering of volcanic eruptions?" mused a renewable-energy expert at a geoengineering workshop in the spring of 2009, to chuckles and murmurs. Robock, who happened to be at the workshop, rolled his eyes. "It's impossible. Nobody knows how to do it. We can't even predict when the next eruption will be or where it will be," he said. "Even after a volcano starts rumbling we can't predict whether it's going to explosively erupt or not." Ken Caldeira grinned. "I think if Edward Teller was around he wouldn't have much of a problem with it," he said.

Getting droplets of sulfuric acid into the stratosphere may even be possible to achieve without accessing the upper atmosphere directly.

An atmospheric scientist named Brian Toon has explored the idea of spraying a chemical called carbonyl sulfide—or COS—at ground level. The planet's natural sulfur cycle would bring the chemical up to the stratosphere, where it would react to form the sulfuric acid droplets. To make the chemical, Toon says, coal burning plants, of all facilities, could be modified so as to emit COS instead of carbon dioxide. The chemical might harm plants, which would slurp it like CO_2 through their leaves, but scientists aren't sure.

In these early days of developing the Pinatubo Option, scientists have imagined creating a haze using something other than droplets of sulfuric acid. "Sulfates are certainly a crude approach," David Keith says. Aerosols of soot are one idea, or dilute potions of custom compounds with names such as boron trihydroxide. Tiny metal particles scatter light much more efficiently than droplets, so their use would require fewer particles to be distributed. Keith has proposed sending up tiny metal disks made of special materials that would rise as sunlight struck them according to the same principles that cause bits of ash released near the surface of Earth to rise, though he's yet to publish the details. "I tell my son that I have actually invented a flying saucer," he says, albeit a minuscule one.

There might be a variety of ways to get sulfate particles into the sky, but scientists who are beginning to study the Pinatubo Option in detail aren't sure how they would behave once there. Modeling performed in 2009 suggested that sulfur dioxide gas sprayed into the stratosphere would form droplets of sulfuric acid that would readily clump. Those larger particles would fall out of the stratosphere before blocking much light. That probably means years of engineering tests—and some believe field tests—would be required to actually make the Pinatubo Option work. (An alternative strategy would be to spray sulfuric acid gas, which some initial work suggests might avoid that problem.)

The Pinatubo Option gets a lot of attention because, in the scheme of things, it wouldn't be very expensive to deploy initially. (Conspiracy theorists who believe in "chemtrails" say the air force is doing it already.) But ultimately, given the stakes, society won't measure the cost of the technique in money—for the same reason that cost wasn't a factor for nations who entered World War II. Intellectual Ventures estimates the cost of its system at $24 million to build and $10 million to operate per year. More complex systems, experts have estimated, would cost in the billions. Getting the haze into place would only be the start. Hacking the stratosphere would, even if carried out as responsibly as possible, require various auxiliary efforts that could raise the total cost into the hundreds of billions of dollars. New monitoring satellites and terrestrial measurement tools would be needed, and, some say, even security forces to protect geoengineering facilities. If and when the world's nations actively take up the geoengineering debate, it's unlikely that costs even in the hundreds of billions would be much of a factor at all, barring some sort of global financial apocalypse. "If I'm going to buy a geoengineering scheme, you can bet I'm going to choose the Rolls-Royce, not the Honda," geochemist Dan Schrag likes to say. If the nations of the world decided to deploy the Pinatubo Option, the climate crisis they faced would be so severe that a cost of several hundred billion dollars would not be a factor.

The real question would be whether nations could bear the long-term environmental impacts and the uncertainty of not quite knowing what the effects of our intervention would be. Acid rain wouldn't be one of them. Sulfur dioxide is a pollutant that comes out of smokestacks and forms acid rain in the lower atmosphere. But as part of the Pinatubo Option, the gas would be released in the upper atmosphere, and even aggressive doses of the approach would add only slightly to the global atmospheric sulfur load. So it wouldn't much worsen the problem.

Yet a thick shroud of doubt envelops the question of how the Pinatubo Option would affect rain. Energy from the Sun drives evaporation, the main source of moisture levels in the atmosphere. And unlike greenhouse gases, which just warm the atmosphere, sunlight warms the atmosphere *and* Earth's surface, where moisture gets turned into vapor. So messing with the amount of sunlight striking Earth could have a greater effect on moisture, scientists think, than shifting levels of greenhouse gases. (Another factor is that accumulating greenhouse gases make it harder for moisture to condense out of the atmosphere.)

When it comes to the effect of aerosols on rain patterns, the clues we have from volcanoes may be misleading. Our climate models are good at the big picture but not the details, where the unexpected effects of geoengineering would manifest themselves. And then there's the frustratingly inconclusive nature of large-scale experiments that involve manipulating the environment. Unless scientists can be absolutely sure that every aspect of the global environment is accounted for in scientific equations, it can be next to impossible to connect X experiment with Y side effect. And it's harder still because scientists would obviously lack a "control" in any geoengineering effort (a planet that geoengineers *didn't* attempt to hack) so as to provide a comparison.

Depending on your perspective, the uncertainty surrounding the Pinatubo Option feels like either an ethical deal breaker or a regrettable price for an idea that might save the human race. Both perspectives were on display at a one-day symposium in Boulder, Colorado, in 2009 to honor climate scientist Tom Wigley, a leading Blue Teamer. Organizers had titled Ken Caldeira's talk "Geoengineering Solutions" on the symposium's program, but he purposely omitted the second word from the title slide of his PowerPoint. "They're not solutions," he told the crowd.

Caldeira's presentation included several maps of Earth in which colored splotches indicated simulated rainfall in a computer model. One map showed a world in which carbon dioxide had skyrocketed to a concentration 44 percent higher than today; another map showed the world with the same carbonaceous atmosphere, except with the Pinatubo Option deployed to counteract the warming. The high-carbon, no-geoengineering world had dark splotches, indicating increased rainfall across Southeast Asia and in parts of

South America and the Pacific. Pink splotches, which indicated less rainfall and likely droughts, festooned the Gulf of Mexico and areas in western Africa. The map indicating the globe with a dose of geoengineering, however, had far fewer splotches of either color. Hacking the planet, in other words, appeared safer than unmitigated global warming.

The question-and-answer session was contentious. “We don’t have good evidence that the precipitation part of this will cancel quite the way you’re describing,” said atmospheric scientist Susan Solomon, who won the National Medal of Science in 1999. She said that observations showing dry years after volcanic eruptions “in the real world” revealed a pattern that “disturbs” her. Caldeira mentioned that other efforts to model the precipitation effects also found that geoengineering actually protected the warming globe from severe disturbances in rainfall patterns. “I think your concerns are valid, but I don’t think we really know the answers yet,” said Caldeira.

“But we know it a lot better than you’re suggesting,” said Kevin Trenberth, another atmospheric scientist. “I wrote a paper about the effects of Pinatubo on the whole of the hydrological cycle.” In the paper, Trenberth had estimated that in the year after the eruption the amount of freshwater dumped into the oceans off the continents each second—a good measure of rain—plummeted by almost 10 percent, a loss of roughly 4 trillion tons.

Volcano eruptions are one-time events, Caldeira responded. Geoengineering, presumably, would be undertaken continually for years or decades, so we don’t know how the atmosphere would respond to a constant dose.

“None of the models do precipitation well enough to do this,” Trenberth said.

“All I’m asking for is a research effort,” said Caldeira.

A separate analysis by Alan Robock, the volcano expert, found that the Pinatubo Option disrupted monsoon patterns over India, “reducing precipitation to the food supply for billions of people.” Monsoon cycles involve parcels of wet air moving inland from the ocean, driven by temperature differences between the land and the sea.

The Pinatubo Option, says Robock, could disrupt the system because as it lowers the total amount of energy Earth receives, the land would cool faster than the ocean, making the two bodies closer in temperature. (The 1991 eruption might have been to blame for the depressed monsoon cycles in Asia that followed, but scientists aren't sure.) Caldeira counters this concern by pointing out that monsoon cycles might be affected differently by a cooling caused by geoengineering because it would continue year after year, perhaps providing sufficient time for ocean and land to equilibrate in temperature. "This is an active area of research," says Caldeira.

Ultimately, of course, the only reliable way to ever know what the effects of the Pinatubo Option will be is to try it. One imagines that if scientists were to try the technique, they'd start small. The problem with small-scale field tests is that they could take twenty-five years for scientists to interpret: were changes in the atmosphere results of the experiment or of natural fluctuations? If geoengineering worked for a small area, could it be scaled up? Computer models would be crucial at answering both questions. In addition, countries would surely demand modeling data that suggested the field tests were safe before supporting any attempts.

The dozen or so world-class models in use around the world have helped scientists answer fundamental questions about the changing climate: Are human activities triggering global warming? ("Greatly," said the IPCC.) How will gross patterns of weather change in a hotter world? (Rainy areas rainier; dry areas drier.) And yet after twenty-five years of developing climate simulations, scientists have little more than an intermediate understanding of how Earth controls its climate, let alone how the greenhouse gases we are spewing forth will change it.

On questions big and small the models fall flat. Since 1979 the fundamental problem has remained the same: insert the same slug of CO_2 into a variety of climate models—say, a doubling of the preindustrial level—and they spit out wildly different pictures of the future. Some say that the globe would warm by 3°F; some say 8°F. The models are even worse on the crucial minutiae, including details directly connected to the frightening questions that hacking the stratosphere raises. Rainfall is a big weakness. The models can't well simulate oscillating seasonal climate phenomena, including El Niño, and predict precipitation for large swaths of the planet with only 50 to 60 percent correlation—scientific jargon for "lousy." They poorly simulate the upper atmosphere, precisely where geoengineers would perform the Pinatubo Option, and may not describe the roles of natural aerosols, let alone billions of tons of new man-made ones. One can't much trust their regional predictions—that is, what the climate will be in twenty years in, say, the Pacific Northwest, or Greece.

This means that Robock's findings about the monsoons may be right, or they may be completely wrong; we don't know. ("He constantly wants to find where the three worst grid boxes on the model output are," Ken Caldeira grumbles about his colleague's predictions.) And yet, despite knowing that they're not very authoritative, it's hard to look at the blue splotches indicating drought over eastern Brazil and Indonesia on geoengineering papers and not for a second feel the *will-this-come-to-pass* shivers.

Perhaps most disturbingly, says German scientist Lennart Bengtsson, the models are much better at simulating the way that "forcings" such as the Sun, carbon dioxide, and various other parts of the atmosphere heat or cool the planet rather than how internal chaos in the system changes things. He's unimpressed with the conclusions about rainfall that early modeling studies like Robock's have made. "It's far too early to be able to rely on these models," Bengtsson says.

Spraying droplets of nearly anything into the upper atmosphere could hamper the recovery of the ozone layer, whose degradation has slowed since the 1990s. The ephemeral layer protects the planet from harmful ultraviolet radiation, which causes cancer. Protecting the ozone layer has prevented roughly 20 million cases of skin cancer, the United Nations says. (Droplets in the high atmosphere—even of water vapor—can harm the ozone by providing a surface on which the reactions that destroy its delicate chemicals can occur.) The 1991 eruption of Pinatubo destroyed some 5 percent of the ozone over the poles and about 2 percent of the ozone over the equator.

But that's for an occasional eruption. How would a sustained geoengineering program affect the planet's delicate yet protective layer? The first time that atmospheric modelers actually modeled the Pinatubo Option to ask that question they returned a mixed verdict. They estimated that geoengineering the atmosphere with 2 million tons of sulfur each year, to balance a doubling of carbon dioxide concentrations, would delay the recovery of the ozone layer by between twenty and thirty years, though they "do not anticipate catastrophic changes" in the stratosphere.

But the scientists acknowledged that the models that come up with these predictions can't really reproduce accurately the almost impossibly complex chemistry that goes on in the atmosphere. And there's the fact that if geoengineers deployed the Pinatubo Option they'd deliver droplets into the stratosphere that themselves would block UV rays, doing the job of the ozone. They might even block enough radiation to make up for the ozone it destroys—provided there's not too much gunk lofted up there. In addition, Crutzen argues that since particles placed in the stratosphere would warm slightly when struck by sunlight, the temperature of the stratosphere would rise. At warmer temperatures, it turns out, the chemical reactions that destroy ozone are less efficient.

The Pinatubo Option might actually deliver benefits beyond cooling the planet. Putting gunk in the atmosphere could help forests grow more abundantly. Blocking sunlight turns out to have a counterintuitively positive effect on the growth of plants for two reasons. First, the particles in the sky reduce the amount of direct sunlight striking the ground, which means that less moisture is taken out of soils. Second, and possibly more importantly, plants use diffuse light—think of gentle mood lighting, provided in the underbrush beneath trees—more efficiently than direct rays. (A 2009 study estimated that urban and industrial pollution from 1960 to 1999 enhanced the ability of forests to take in carbon by a whopping 25 percent.) The Pinatubo eruption tripled the amount of diffuse light striking Earth's surface, and scientists calculated that the change was responsible for speeding photosynthesis by 10 to 20 percent in one deciduous forest. As Earth's forests were enhanced by the diffuse light, they sucked in more carbon dioxide each year, slowing the rise in the carbon concentration in the atmosphere in the two years following the eruption.

And yet ecologist Tony Janetos of the Pacific Northwest National Laboratory says the overall ecological impacts of the Pinatubo Option will be extremely hard to predict. If the technique jiggers global rainfall patterns, the effect on ecosystems could be deadly. Scattering sunlight might not improve the ability of biomass to grow year after year, despite the effect that was seen after a one-time event such as Pinatubo. After all, scientists have enough trouble trying to understand how the gradual effects of global warming are altering a variety of ecosystems right now. So asking them to predict the effects on ecosystems of geoengineering—which would affect in complex ways both sunlight and carbon dioxide, the two main elements of plants' diets—is just too much to ask right now, he says. In the past, ecosystem models have responded to simulated climate change "essentially identically when running on today's climate." But when scientists have simulated a future climate with high carbon dioxide levels, the responses of the models have been all over the map. "Some models greened up dramatically, some models didn't green up at all," he said. So when it comes to predicting how geoengineering might affect ecosystems, "we are really on uncharted ground."

How the Pinatubo Option might affect our ability to collect energy from sunlight is easier to predict. Dirty skies have thwarted solar power facilities in the past by converting direct sunlight into diffuse light by scattering the sun's rays. By the same token, the Pinatubo Option, in a perverse twist, could actually harm efforts to produce carbon-free power. Solar panels would be mostly unaffected because they principally rely on diffuse light. But another kind of solar power involves a power plant using thousands of mirrors reflecting direct sunlight to a white-hot power station that converts heat to electricity. In 1982, dust from the eruption of a Mexican volcano slashed by a quarter the amount of produced power from one such plant in the Mojave Desert. ("The engineers didn't know what was going on," recalls scientist Michael MacCracken of the Climate Institute in Washington D.C. "They were using a three-hundred-dollar detector. When they switched to the six-hundred-dollar model, that measured the direct radiation from the Sun, they realized what was happening.")

Solar power stations that use curved mirrors to track the rays of the Sun, known as solar thermal facilities, also rely on direct sunlight. At one such facility, also found in the Mojave Desert, the Pinatubo eruption in 1991 reduced the total amount of solar energy striking the surface of the mirrors by 3 percent. An even bigger effect was a big increase in diffuse light, which dropped annual solar output by 14 percent. "That the power output of those plants dropped by so much after Mount Pinatubo when we all know the sky didn't go dark in 1991 is, I think, surprising until you think about it," says atmospheric scientist Dan Murphy.

At the National Academy meeting on geoengineering in 2009, federal research manager Joel Levy listened in increasing horror to various proposals to block sunlight and cool Earth. Early in his career he had done research on solar panels at MIT. On a paper in front of him he wrote, "Dr. Strangelove stuff." He stood. "The origin of our climate change problem is our failure to utilize solar energy," he said. "What we are actually talking about here is throwing away what we should be harvesting." He looked around the room. "We've stepped right through the looking glass."

The Pinatubo Option tends to inspire strong emotions. Trenberth—a committed Red Teamer—agreed to review a paper Wigley wrote in 2006 for *Science* that proposed simultaneous emissions cuts and geoengineering. But Trenberth told Wigley as well as *Science*'s editor that the paper's failure to discuss the risks of drying the planet out was a serious omission. When he felt they ignored his point, he angrily submitted a letter stating that the Pinatubo Option would create "a risk of global drought" while failing to address the underlying problem: "like telling a patient with a broken arm to take two aspirin to cope with the pain." Then, in a short comment in *Physics Today,* he included the following fable:

> Once upon a time in an idyllic country, near a small town and a farming community, a rope hung out of the sky. One pull on the rope changed the weather from fine and sunny to cloudy and rainy, and the next pull changed it back. For many years the people cooperated; the farmers used the rains to help grow crops, and the townspeople enjoyed the sunny periods. But there came a time when the townspeople protested the rain and wanted more sunshine. The farmers were concerned about their crops. And so arguments broke out, with a person from the town pulling on the rope, followed quickly by a farmer pulling it again, and they pulled and pulled and . . . broke the rope.

"Ethical considerations," he wrote, required that forms of climate manipulation that propose to block sunlight "would be simply unacceptable."

"On the whole, the world's precipitation patterns would be less affected under a geoengineering scenario than in a high-carbon world, but I'm willing to acknowledge there might be some places that have very adverse effects," Caldeira told me later. "That's why this is an emergency technique. No politician in their right mind is going to want to deploy this unless they have to."

But even during an absolute crisis, what would they know about the side effects? Princeton University scientist Michael Oppenheimer says the idea of using geoengineering in an emergency is a "red herring" because it assumes, in a contradictory way, that future scientists would have sufficient information to act if need be. "If climate change leads to such surprising outcomes that we are caught unaware and need to do something quickly, then that means our models would be useless for projecting the consequences of implementing a geoengineering option."

So, if Oppenheimer is right, the nations of the world would only hack the stratosphere in a worst-case scenario; but in a worst-case scenario, by definition, they'd be flying blind. But it's worse than that. The world also might become addicted. Deploying the Pinatubo Option on a global scale could encourage some to continue burning fossil fuels because they considered the problem solved. The level of carbon dioxide would rise in the atmosphere, while the temperature of the planet remained constant or fell. This would create an obligation to geoengineer in perpetuity to keep temperatures low. The planet's carbon dependency would become a geoengineering habit we couldn't break. "Employing geoengineering schemes with continued carbon emissions could lead to severe risks," a scientist named H. Damon Matthews wrote in a paper with Ken Caldeira.

Say for one reason or another—for example, political disruption or terrorism—we stop geoengineering. The results could be much more devastating than the current warming that the planet is experiencing. In 2006, wearing their Red Team caps, Matthews and Caldeira ran a supercomputer simulation they named OFF_2075. In that simulation, the Pinatubo Option was used to keep temperatures down while nations continued to burn fossil fuels until the year 2075, when the geoengineering halted abruptly. Pushed by the accumulated greenhouse gases in the atmosphere, temperatures spiked at the rate of 5°F per decade—twenty times faster than the current rate of warming. In the past hundred thousand years, the planet has never warmed up so quickly. Halting the scheme, the pair wrote, could cause a "warming rebound" that could cause "large and rapid temperature oscillations" and "severe impacts on both human and environmental systems" that scientists could only imagine.

"Accumulated pent-up climate change would be unleashed upon the Earth," Caldeira said later. The scene, he said, would be of the "dystopic world" in the 1983 film *The Day After*, about the apocalypse that follows the aftermath of a nuclear war on a small Kansas town.

IN 1997, ALGAE HAD TURNED THE SURFACE OF SEVERAL LAKES IN Queensland, Australia, into a light green soup. Stocking the lakes was a regular practice for the enjoyment of recreational fishermen. But Vladimir Matveev, an ecologist with the Australian government, thought that adding an extra large amount of big fish to the lakes might help control the ugly blooms.

He was following a theory known to be operative in North American lakes. The large fish ate the small fish, which ate the lake's crustaceans, which ate the algae. By removing the small fish from the ecosystem, Matveev figured, their prey, the crustaceans, would grow in numbers. More crustaceans, also known as "micrograzers," would mean less algae. The ecologist arranged a five-year experiment in which he added three times the usual amount of Australian bass to one of the lakes, Lake Maroon. After the first year, its ecosystem behaved as he hoped, and the waters remained free of blooms of algae.

"We were about to celebrate our success when the population of effective micrograzers crashed," he said. "Then, as stocking continued, it was a complete reversal of the situation, and a potentially toxic soup of [algae] exploded across the lake." The lake took on a deep green sheen. Later, Matveev added fewer large fish to the lake, and it recovered. Matveev believes that there is some sort of limit after which adding big fish stops having a beneficial effect on the food chain. But ecologists haven't yet understood the dynamics at play.

The Pursuit of Levers

Can climate science exist without the allure of climate control? Each generation of weathermen has had its rainmakers. In 1896 Swedish physicist Svante Arrhenius became the first to suggest that increasing CO_2 might raise the planet's temperature substantially. He urged humanity to deliberately do so by burning coal, envisioning warmer climes. Respected meteorologist James Pollard Espy, America's first federal weather scientist, risked his reputation in 1845 with a nearly pyromaniacal desire to burn forests in pursuit of artificial rain. More than a century later, a prescient U.S. government report for President Lyndon Johnson calling carbon dioxide "the invisible pollutant" offered geoengineering as its lone climate solution.

The fits-and-starts evolution of planethacking, as a global endeavor, and weather modification, its more localized cousin, has paralleled the spurts of progress in fundamental climate science. And those spurts have in large part been driven by scientists' search for levers, the small changes in Earth's system that can have profound global effects.

Nineteenth-century scientists saw the severity and pattern of local weather as predictable and implacable, except over very long periods. So when they devised ways to control rain they did it through the application of the explosive force they thought was required. Similarly, when the Earth's global climate appeared to be a static, stable, relatively unchanging system, experts dreamed up massive ways to control it to reshape nature so as to serve their needs. They imagined rechanneling rivers, shifting ocean patterns, or launching giant space mirrors.

Later in the twentieth century, climate scientists realized, however, that the atmosphere was much more sensitive to small changes than they had known. And so, as the Anthropocene began, they learned both how humans inadvertently were shaping climate and how geoengineers might one day shift the global climate on purpose. They utilized—or fantasized about utilizing—levers that had been discovered before. Geoengineers of the early twentieth century, contemplating brute force, provided today's atmospheric interventionists with their sense of global aspiration and scale. The atmospheric scientists who later discovered the levers bestowed unto the nascent field of geoengineering an appreciation for just how sensitive the atmosphere is—and where to push.

At the end of the nineteenth century, the idea that small prods might spur profound and rapid changes was a deeply foreign idea to mainstream Earth science. Geology had been a battlefield won by uniformitarianism—the belief that the forces shaping the Earth had been relatively slow and gradual. The loser was catastrophism, the religiously tinged idea that the Earth's history had been punctuated by quick, violent upheavals. (Plate tectonics, which postulated flowing continents and an ever-changing global map, wouldn't be proposed until 1915 and accepted as fact until the 1960s.) By the same token, scientists believed that the atmosphere behaved according to set physical laws like the other parts of nature. That helped explain what they thought to be its consistent behavior over time. What they underestimated was the complexity of the atmosphere. Later, scientists would realize it wasn't just physics that determined its behavior.

Scientists of the period who sought to bring forth rain used violence. Espy, born in 1785, had won acclaim from his peers by establishing a theory that storms resulted from heat flows through the sky, and he became the first weather scientist to receive U.S. government funds. But the newspaper headlines he garnered heralded instead his attempts to generate artificial rain by fire. "In the summer of 1849 he contracted for 12 acres of timber in Fairfax County Virginia 'with pines as thick as a man's leg or arm,'" wrote weather historian James Fleming in an essay titled "The Pathological History of Weather and Climate Modification." The experiment failed, he wrote. A retired Civil War general named Edward Powers claimed in 1871 that battle records from that conflict suggested that smoky artillery engagements prefaced rainfall, and Congress gave him $2,500 to test the idea. "Most likely there was no correlation between battles and storms, but generals chose to fight during breaks in the weather," wrote Fleming. And yet former general Daniel Ruggles received a patent and federal money to conduct "what one observer called 'a beautiful imitation of a battle,'" Fleming writes, armed at the site of the experiment with "an arsenal of explosives, including balloons and kites to be detonated at various altitudes."

A number of pioneers throughout the recent history of atmospheric science have sought to control the climate as they unraveled its secrets, including some of the field's earliest pioneers, such as Arrhenius. The precipitously rising concentration of carbon dioxide in the atmosphere may today cause alarm, but it's rarely emphasized how small a fraction it is: fewer than 4 molecules per 10,000. In 1894, when Arrhenius's wife, Sofia, left the portly Swede, there were fewer than 3 molecules of it per 10,000. Scientists at the time had no idea that subtle changes in atmospheric trace gases could cause big changes in global temperature. They instead blamed what little changes they detected on the movements of the planet itself around the Sun, which were believed to control the occurrence of ice ages.

But intrigued by a colleague's views on the role of volcanic gas to alter the planet's temperature, Arrhenius wasn't sure. Despite his despair—Sofia wouldn't let him see their newborn child, and taunted him in letters with reports of her happiness—the Swede plunged into an intense effort to create what was essentially a global energy map on paper. He sectioned off the Earth into small boxes to calculate local heat budgets, working for months in the nearly eighteen-hour-long sunshine of the Stockholm summer.

Cutting the number of carbon dioxide molecules per 10,000 from 3 to 2, Arrhenius calculated, would trigger another ice age. Conversely, he calculated that if the fraction of the gas in the atmosphere were to double, average temperatures would skyrocket by 10°F. It was this minute concentration of CO_2, Arrhenius believed, that underpinned his theory of why Earth oscillated between what he called "glacial" and "genial" periods.

Arrhenius not only observed the effects humans could have on the atmosphere, he also promoted those effects aggressively. In his 1908 book *Worlds in the Making*, he encouraged humanity to pursue a "virtuous" course in which the continued burning of fossil fuels staved off a future ice age. Furthermore, by improving the global temperature, local climates, such as Sweden's, could help grow "much more abundant crops than at present, for the benefit of rapidly propagating mankind." Nils Ekholm, a Swedish meteorologist, suggested several ways to pull this potent lever. In 1901 he proposed that mankind could turn up the thermostat by digging "deep fountains," presumably to tap natural CO_2 vents. Burning shallow coal beds was another means of adding "carbonic acid," as carbon dioxide was called at the time, to the atmosphere. Or, he suggested, society could control the level of greenhouse gases in the atmosphere by varying "the growth of plants according to [its] wants and purposes." He wrote in 1901:

> It seems possible that man will be able efficaciously to regulate the future climate of the Earth and consequently prevent the arrival of a new Ice Age. . . . It is too early to judge of how far men might be capable of thus regulating the future climate. But already the view of such a possibility seems to

me so grand that I cannot help thinking that it will afford to Mankind hitherto unforeseen means of evolution.

It would be four decades before scientists were to understand how sensitive the Earth system was to minute changes in the concentration of this trace gas or temperature. Arrhenius believed Earth could warm by as much as 10°F in several thousand years, which implied that an ice age could come about just as quickly. "That was much faster than anybody thought there could be an ice age. People didn't take it seriously," historian Spencer Weart told me.

Among other reasons why early scientists downplayed the significance of carbon dioxide, experts generally believed that atmospheric CO_2 already absorbed all the reflected solar radiation reflected up from Earth, so adding more wouldn't increase the greenhouse effect. When climatologist C. E. P. Brooks suggested in 1925 that slight changes in conditions could cause abrupt or even catastrophic changes—by exposing dark ground, which could melt ice sheets in perhaps "a single season"—scientists found the concept "preposterous," Weart wrote. "Among other arguments, they pointed out that ice sheets kilometers thick must require at least several thousand years to build up or melt away."

Flawed geological data and bad analysis cast further doubt on the idea that small changes could have large, rapid effects. In the first half of the twentieth century, oceanographers relied on temperature measurements inferred from silt and clay samples recovered from the floor of the ocean. They didn't show rapid changes, but the scientists didn't know how flawed these data, known as temperature proxies, were. (Crawling worms mucked up sediments, obscuring the records.) Clay found in ancient lakebeds did show rapid temperature spikes, however, as did new kinds of analyses of ancient pollen found in 1955. But scientists dismissed both clues as evidence of local changes, not global cold snaps.

• • •

With the dogma of an unchanging and unyielding Earth as the backdrop, the nascent superpowers of the twentieth century formed plans to remake nature to suit their national missions. The American effort to reshape its frontier was realized most officially in the 1902 establishment of the U.S. Bureau of Reclamation by Theodore Roosevelt. Its objective was to irrigate the American West. The agency would tackle seventy water projects between 1902 and 1944, including construction of the Hoover Dam and large-scale water projects in Missouri, Colorado, and Washington State.

Physically remaking their country took on a more systematic and grandiose effort for scientists in Russia, and later in the Soviet Union. Surviving the climate's harsh conditions had always been part of Russian culture; Soviet ideology and modern science merely provided new tools. "We have a saying in Russian: we cannot wait for charity from nature, our task is to take it," Russian scientist Valentin Meleshko said recently.

Some of the nation's most cherished scientific forefathers were instrumental in the task of taming the unforgiving land, including soil scientist Vasily Vasil'evich Dokuchaev. In the 1890s Dokuchaev was asked by the tsar to tackle the problem of frequent droughts in southern Russia, where farms were languishing in the dry conditions. An extensive array of trees and ponds that he ordered were installed successfully. They changed local hydrological conditions by trapping snow and adding moisture from evaporation to the air around the farms.

Geochemist and mineralogist Vladimir Ivanovich Vernadsky later provided a scientific framework for the Soviet attitude regarding humanity's true relationship with nature. His book, *The Biosphere*, published in 1926, attempted to smash the boundaries between the inanimate and the living segments of Earth. Life itself was a geological force, he wrote, and humans had a special role. Within his integrated framework of connected sections, the "geosphere" contained earth, rock, and water; the "biosphere," living creatures. The noosphere—from the Greek *noos*, for mind, and coined, not surprisingly, by a French philosopher—included humans as well as their technology. Human lives and progress were essentially the "accumulation and transformation of the luminous energy of the Sun," Vernadsky wrote, just like the activities of other organisms. "As Darwin showed all life descended from a remote ancestor, so Vernadsky showed all life inhabited a materially unified place," wrote scientists Lynn Margolis and Dorian Sagan. "Life was a single entity transforming to earthly matter the cosmic energies of the sun."

Galvanized by the idea that remaking the Russian land was the natural role of the Soviet people, Stalin called on his nation to transform their harsh environs with the same vigor they had marshaled against the Nazis. A 1948 plan passed by the Soviet Communist Party detailed the new direction for dramatically reshaping the 8-million-square-mile nation. "Within days, a schematic map of the European USSR appeared in kiosks throughout [Moscow]," wrote Paul Josephson and Thomas Zeller in 2003:

> According to the fantastic map, all major rivers had been dammed; a vast irrigation system spread into fertile but arid land of the Southern steppe; hydroelectric power stations were distributed liberally; huge reservoirs filled up behind them; canals and locks guaranteed ease of inland water transport; scores of "forest belts," dozens of kilometers long and several hundred meters wide, ringed land that has been plagued by constant dry winds but now would be fertile and lush. Nature, like the Russian peasantry, bourgeoisie, and society itself, would succumb to the Party's will.

Soviet atmospheric scientists followed suit. Northern hemisphere temperatures were falling slightly in the middle of the century, just as climate scientists around the world were learning how quickly the planet had moved into glacial cycles in the past.

The most prominent climatologist in the Soviet Union, Mikhail Budyko, was to become the founding member of the geoengineering

Blue Team. He used a simple climate model to suggest that a slight weakening of the greenhouse effect could lead to a completely iced-over "snowball Earth." As snow and ice advanced toward the equator, he showed, the planet would cool while reflecting solar energy away in a feedback loop. The concern about an increasingly fickle climate that might be getting cooler coincided with Soviet interest in possibly opening up the Arctic for shipping. Budyko patented a method of distributing industrial soot over frozen areas to rapidly melt ice there. "For example, from the shoe industry," says Andrei Lapenis, a former student of his. "People would say when I saw him, 'That's the guy who proposed to melt ice with ash.'" Another method Budyko proposed was to spread chemicals on the Arctic Ocean to change its physical properties.

Others were far more daring. A pair of Russian engineers named Gorodsky and Cherenkov proposed alternative projects that would create dust rings around Earth, à la Saturn, which would reflect additional solar energy to Earth. Gorodsky's version would be "shaped like a flat washer whose lower boundary would be 1,200 kilometers from the surface of Earth, with its upper boundary line at an altitude of 10,000 kilometers," wrote Nikolai Rusin and Liya Flit in a 1960 propaganda pamphlet, "Man vs. Climate." Russian atmospheric scientists dreamed up ways to control rain, snow, fog, and hail, they wrote. "We are merely on the threshold of the conquest of nature."

Diverting Siberian rivers was one way Rusin and his colleagues tried to further that effort. An experiment in 1960 used cloud seeding techniques to clear the sky over a Russian region of nearly 8,000 square miles. A scientist named Petr Mikhailovich Borisov, meanwhile, proposed building a 74-kilometer-long dam across the Bering Strait to pump cold Arctic water to the Pacific, pulling warmer Atlantic water into the Arctic basin on the other side. Borisov said the project would warm northern Asia by more than 30°C, melting the permafrost, turning tundra into "regions of large-scale cattle raising," as he said in 1969, and bringing grass to the Sahara. (President John F. Kennedy was said to have considered joint work with the Soviets to study Arctic-melting ideas.) The demise of Greenland as it melted would raise global seas slowly, Borisov wrote, but the "very attractive" advantages were worth it.

It was a radical vision of a remade world. "The temperature of the air and water on our planet will undoubtedly become higher and more uniform. The sharp contrast between the north and south will disappear, the polar ice will melt," wrote Rusin and Flit. "If we want to improve our planet and make it more suitable for life, we must alter its climate. Just as today we plan the construction of new cities, creation of new seeds, conquest of space, so in the future we shall have to plan improvements in the climate."

Stanford climate scientist Stephen Schneider calls Rusin and Flit's work an "upbeat little pamphlet" and "entertaining." But, he says, "I'm glad they failed."

The Soviet plan to physically transform their climate certainly deserves the "geoengineering" moniker by virtue of its scale and ambition. In the United States, a confluence of factors drove what was to be twenty-five years of intense weather modification research. First was a growing, if still rudimentary, understanding of the subtle phenomena that controlled the atmosphere, starting with the discovery of cloud seeding at General Electric's research labs just after World War II ended. Then there was the advent of computers. They provided influential researchers such as mathematician Johnny von Neumann a tool with which they could imagine, if not gain, unfettered control of perhaps the most complex system on the planet. And finally there was the Cold War itself, with all its fear and ignorance, and the role of researchers in waging it. Influential scientists openly feared that the USSR's atmospheric scientists were opening up a new front in the sky.

Over the course of a single month in 1946, two GE scientists discovered how the addition of tiny particles could change the number of water droplets within clouds, causing rain or snow. "A single pellet of dry ice, about the size of a pea," wrote the *New York Times*, might produce enough ice nuclei to develop several tons of snow." GE research manager and chemist Irving Langmuir, a Nobel Prize winner, believed "in general all meteorologists need was to find a proper trigger to release the immense amount of energy stored in the atmosphere." Historian James Fleming points out the nuclear analogies: "cloud seeding nuclei, cause chain reactions forming a destructive mushroom shaped cumulonimbus cloud," he writes. Langmuir claimed the technique had altered the direction of a hurricane and caused 8 inches of snowfall in separate field trials. California Institute of Technology meteorologist Irving Krick translated the idea of atmospheric leverage into monetary terms, claiming in 1949 that an investment of $100,000 in silver iodide generators on the ground would yield rainfall worth $14 million in water.

As the modern age of weather modification began, it shared not only a scientific vocabulary with the budding atomic revolution of the 1950s but also its technological enthusiasm. It was the golden era of American big science; nothing was bigger than the sky. Generals and businessmen took to the new technology with equal verve. Starting in 1947, Langmuir's GE scientists partnered with the Pentagon on a five-year program involving 180 field experiments. The nation "that looks to control the time and place of precipitation will dominate the globe," Strategic Air Command commander George Kenney said that year. The navy's Project Scud involved feeding cyclones; the air force seeded cumulus clouds with water delivered via B-17 bomber; the weather bureau looked into self-propagating rainstorms. But the experiments were such a failure, wrote Fleming, that weather scientists went back to the drawing board for fundamental studies after the effort was completed. The stakes were high, or so it seemed: *Newsweek* declared a "new race with the Reds" to develop the "weather weapon." By 1960 annual federal support for weather and climate modification rose to roughly $10 million.

Atmospheric scientists proposed a techno-fix for every problem. As smog enveloped Los Angeles in the mid-1950s, experts suggested a bevy of solutions: deploying hovering helicopters to blow the dirty air downward and fans to blow it upward, painted roofs in a checkerboard pattern to disrupt air movement, or blasting the smog with warm air produced by burning garbage beneath two-thousand-foot-tall smokestacks. A better solution than any of them, wrote meteorologist Morris Neiburger in 1957, would be to simply stop polluting. The multi-agency Project STORMFURY sought to weaken Atlantic hurricanes by seeding them. But it failed to yield statistically meaningful results despite the involvement of hundreds of scientists over the program's duration between 1962 and 1983.

That a small amount of seeding particles could produce explosive results made the venture incredibly alluring for enterprising businessmen, despite the fact that scientists didn't understand what they were doing (or, for that matter, the possible side effects, such as seeded clouds robbing moisture from nearby areas.) But fledgling commercial weather modification firms made brisk business in the postwar era. By 1951 the young industry earned more than $3 million per year and had targeted almost 15 percent of the area of the lower forty-eight states. It was an enterprise fraught with legal entanglements, however, though scientific uncertainty involved made it difficult for lawyers to prove anything in a series of cases in which litigants alleged water stolen or floods caused. Scientists were hauled into court, and the U.S. government began to sour on weather modification research after a 1972 government rainmaking experiment was followed by a flood that killed 283 people in Rapid City, South Dakota. (After an investigation, the U.S. Bureau of Reclamation ruled there was no link between the research and the disaster.)

The military eventually began deploying its weather weapons, with decidedly mixed results. A 1965 effort to burn a forest in which Viet Cong were hiding, for example, went awry after the blaze was ignited. "[T]he 'thermal convective condition,' as U.S. Air Force meteorologists later called it, triggered a drenching downpour that doused the forest fire and left Boiloi's Viet Cong safe and unsinged in their caves," reported *Time* magazine. Six years later, the press learned of Operation Popeye, a secret air force program that flew twenty-six hundred sorties over five years in an attempt to create rain over the Ho Chi Minh Trail. Scientists were aghast, the Senate passed a resolution against weather war, and the Soviets, "taking full measure of the Watergate crisis, seized the diplomatic initiative" by unilaterally bringing the issue of weather modification to the United Nations, wrote Fleming. Signed by seventy-three nations, the 1976 Convention on the Prohibition of Military or Any Other Hostile Uses of Environmental Modification Techniques banned belligerent rainmaking.

For the most part, cloud seeding remains today a large-scale though scientifically controversial endeavor, hidden in plain sight. One expert estimates that up to $100 million are spent on efforts in more than thirty-five countries, including projects that aim to increase the snowpack in the Rocky Mountains and Sierra Nevada and increase rainfall in South and North Dakota, Texas, and Florida. It's hard to prove that seeding schemes work the way they're advertised; it's hard to prove to their fans that they don't. A panel of experts convened by the National Research Council in 2003 concluded that although weather modification experiments are a "legitimate element of atmospheric and environmental science," they lack the scientific basis that would allow them to deliver "predictable, detectable and verifiable results."

While would-be rainmakers in the middle of the twentieth century sought out local levers in the sky, climate scientists were beginning to understand that the planet's atmosphere was sensitive to small changes on a global scale.

In 1924 a Serbian physicist named Milutin Milankovic had determined how minute wobbles in Earth's orbit, tilt, and spin subtly altered the pattern of sunshine striking the planet. So-called Milankovic cycles, as they were to be called, delivered slow shifts in the tilt of Earth's axis, resulting in 100,000-year ice age cycles. The changes he imagined were slow and steady, fitting with the conception of a slowly changing Earth that dominated the first half of the twentieth century. "Scientists had rejected old tales of world catastrophe, and were convinced that global climate could change only gradually," wrote climate historian Spencer Weart.

But in the late 1950s scientists began to suspect that other phenomena could augment such astronomical forces to make the planet's climate lurch capriciously. At Columbia University in 1956, an analysis of ancient seafloor shells collected from cores drilled by scientists suggested that temperatures had risen about 2°F per thousand years about 11,000 years ago. That conflicted with "the usual view of gradual change," admitted one of the scientists at the time. Five years later, meteorologist Reginald Sutcliffe employed the new science of systems controls, known as cybernetics, to describe the world's climate as a complex system in which "sudden jumps" were inherent features. Also in 1961, using a simple model of heat flows, Budyko, the Soviet scientist, declared that "the Arctic ice pack might disappear quickly if something temporally perturbed the heat balance," wrote Weart.

Scientists' evolving interest in exploiting atmospheric levers for human ends was not confined to this planet's skies. A twenty-seven-year-old Carl Sagan, relatively unknown at the time, utilized the concept in 1961 to envision reshaping the atmosphere of Venus in what was a key paper in a kooky quasi-discipline known as terraforming. The temperature at the planet's surface is roughly 900°F. Toward the conclusion of an otherwise unassuming review in *Science* of research on the planet, Sagan suggested spraying the Venusian atmosphere with algae to cool it down. The bugs would devour the carbon dioxide in Venus's sky, he reasoned, reducing the greenhouse effect there. That would tip the temperature of the planet below the boiling point of water, creating a runaway cooling effect. Sagan figured that the moisture that then condensed in the atmosphere would catalyze chemical processes on the ground to suck even more carbon dioxide out of the Venutian sky, further and further cooling the planet down. (Twelve years later, Sagan would suggest that decreasing the brightness of the Martian polar ice cap by "only a few percent" with black carbon dust might trigger beneficial warming there, as well.)

Perhaps unsurprisingly, given optimistic attitudes toward technology that pervaded at the time, geoengineering received top billing in a government report on climate in 1965. The authors were a panel of distinguished researchers appointed by President Lyndon Johnson's scientific advisory council to broadly assess "the quality of the environment in which our people live." The threat of man-made carbon dioxide—or as the report put it, "the invisible pollutant"—received an entire chapter. It was a groundbreaking document—the first U.S. government report to thoroughly examine the problem of global warming. "Deleterious" effects, it said, could include rising sea level or disrupted fisheries. The panel's proposed solution? "The possibilities of deliberately bringing about countervailing climatic changes . . . need to be thoroughly explored," said the report. Spreading reflective particles on the surface of the oceans was raised as a possibility, or conducting seeding operations at high altitudes to form cirrus clouds. No mention was made of reducing greenhouse gas emissions.

The more scientists learned about the climate system, the more unstable and responsive to small changes it appeared to be. In the same year as the report to Johnson, mathematician and weather scientist Edward Lorenz, using a simple computer model, found that tiny changes in the climate system might trigger huge changes by virtue of the inherent unpredictability of the interwoven system. Lorenz had invented chaos theory. The following year scientists from Columbia University accumulated ancient evidence of an "abrupt transition" between so-called glacial and interglacial states 11,000 years ago.

Then, in 1969, Budyko used a simple heat model derived from eighty years of temperature data to suggest that the planet's climate was so precarious that as little as a 1 percent decrease in solar radiation could cause an ice age. And that change could be caused by a large volcanic eruption. But, he wrote, the carbon dioxide that humanity was putting into the atmosphere was having the opposite effect, and could forestall the "coming climatic catastrophe" of global heat waves or a snowball Earth. Soon after, a scientist in Arizona, using different equations than Budyko, assigned Earth the same unstable behavioral disorder. "We didn't know at the time how sensitive the system was, or that changes could happen so rapidly," recalls Stanford climate scientist Stephen Schneider. "That work started getting people concerned. We said, 'Wait a minute, this isn't just a matter of small changes—we could see big changes.'"

The possibility of runaway change had a particular effect on the way Budyko viewed his nation's relationship with the frigid Arctic. Once among the scientists who had imagined melting the frigid expanse, the Soviet patriot changed his opinion, questioning in a 1971 book whether purposely destroying ice was "suitable" given the vagaries of the global climate system, and the risks that might be inherent. Instead, he suggested, "It will be necessary to work out methods, not for the destruction of polar ice, but for its temporary or long-term preservation." A year later, dismayed that rising greenhouse gases might bring rising seas and other ecological calamities, he proposed blocking the sun by lofting droplets of sulfates into the high atmosphere using modified jet fuel. More than three decades later, Paul Crutzen would champion the Pinatubo Option as a dire course of action humanity was now forced to consider.

Other atmospheric levers were even easier to pull inadvertently than those that controlled global temperature. In 1974 a pair of scientists theorized that the refrigerant Freon—called "one of the most outstanding scientific achievements of our times" by Frigidaire when it was introduced four decades before—was destroying the ozone layer. They suggested that once Freon or other similar refrigerant gases arrived at the upper reaches of the sky, ultraviolet light pried loose from them a chlorine atom. It turned out, scientists soon learned, that each chlorine atom destroyed 100,000 atoms of ozone. The chlorine was a catalyst. It could be used over and over. The power of the chemical's effect on the atmosphere startled atmospheric scientists, including meteorologists who even as late as the 1970s "thought of the stratosphere as this unknown land," says Ralph Cicerone, an atmospheric scientist who today heads the National Academy of Sciences. "It's not that you're trying to put a two-by-four and move the whole planet as a lever," Cicerone told me. "But moving one small part of the system can have a planetary impact."

On May 15, 1989, sitting in a plane waiting to depart the Moscow airport, Cicerone came up with a scheme to fix the problem of the destructive chlorine by chemically binding it with another gas in the stratosphere. Two years later he and colleagues published a paper in *Science* in which they mused that "a fleet of several hundred large airplanes" might solve the problem by spraying 50,000 tons of ethane or propane gas over Antarctica, though they acknowledged pressing questions about side effects or related legal and ethical issues. "The scientific community was not very happy with that paper," Cicerone says. A few years later chemists showed the scheme couldn't work.

The closer scientists have looked at Earth's moving parts, the more levers they have uncovered. In 1986, chemist James Lovelock was standing in the University of Washington office of Robert Charlson. They were trying to figure out how droplets of an organic chemical that algae breathed into the atmosphere might affect Earth's temperature. "What if there's an amplifier in the system?" wondered Lovelock. Charlson, an expert on clouds, pulled a textbook on atmospheric aerosols down off the shelf and turned to page 289. There a graph showed that adding droplets to clouds would have little effect on their brightness, until, with enough droplets added, suddenly their reflectivity would spike up. "Good heavens, Bob," Charlson recalls Lovelock saying. "That's the amplifier we need!"

The following year the pair and two others published an extremely influential paper suggesting that tiny amounts of the chemical, known as DMS, could have global effects by exploiting the "amplifier" that the properties of clouds provided. Small changes, in other words, worked to keep the planet's temperature stable. When there were fewer clouds, less sunlight would get reflected into space, so more would strike the surface of the ocean and warm it up. The extra sunlight would meanwhile enhance the algae's growth, but as the algae grew they would produce DMS, which would enhance the clouds' brightness. To close the loop, the scientists figured, more clouds would mean less sunlight hitting the ocean, diminishing algae growth and providing a feedback in the other direction.

In 1990 another scientist, Anthony Slingo, quantified the levers, suggesting that "modest" changes of roughly 20 percent in the number of clouds or their moisture would be sufficient to counteract the warming effect caused by a doubling of the concentration of atmospheric CO_2. Soon after that, the idea inspired British cloud scientist John Latham, who that year proposed the geoengineering method of brightening clouds at sea with salt water to cool the planet.

One day in 1977 the phone rang at the desk of Mike MacCracken. He was head of atmospheric sciences at Lawrence Livermore National Laboratory, a nuclear weapons lab outside San Francisco. It was Genevieve Phillips, personal secretary to chemist Edward Teller, the father of the H-bomb, and the scientific and ideological leader of the controversial laboratory. A future proponent of geoengineering had a scientific problem for MacCracken.

"Mike, he wants to see you in two hours," Phillips said to MacCracken.

"Oh, what now?" MacCracken asked.

"One of his people is asking the question 'Could we use nuclear explosions to break the California drought?' and he wants to talk to you about that."

MacCracken laughed to himself. "Okay," he said, and hung up.

When MacCracken showed up at Teller's office two hours later, he had no idea how one might actually use atomic blasts to tackle the problem. But he had roughly calculated the total energy that would be required. "Teller's style was 'get to the blackboard and convince me,'" MacCracken recalled. Grabbing the chalk, he first estimated the rough amount of energy released upon condensation of the state's annual rainfall: roughly 1 trillion billion calories. One theory as to why the drought was so severe was that rainfall was being diverted as a result of a particular pattern of ocean temperatures out at sea. Another was that excess snowfall in the Great Plains was sustaining regional cooling and affecting circulation patterns. So MacCracken calculated the energy involved in the ocean patterns and the snow and got a similarly gargantuan amount of energy. Translating that amount into explosive force, MacCracken estimated that engineers would have to deploy roughly a million 1-megaton bombs. No levers, in other words, were available.

"That was the end of that idea," MacCracken recalled, as Teller was convinced to let the concept die on the radioactive vine. MacCracken went back to his office, where Teller had assembled one of the best atmospheric modeling outfits in the world, and went back to work. MacCracken laughs when he recalls the story. "Teller was an order of magnitude thinker," he says.

As much as for any twentieth-century scientist, Teller's life was devoted to seeking technological solutions for humanity's problems—a pioneer who searched for scientific levers and ways to utilize them. Among Teller's last scientific publications was a 2002 technical paper titled "Active Climate Stabilization," in which he called for the Pinatubo Option instead of carbon emissions cuts. (He died at age ninety-five a year later.)

Livermore's explosive relationship with the planet and Teller's unappreciated role in encouraging atmospheric research help explain how he came to promote geoengineering as one of his final public acts. During the Cold War and its aftermath, the low assemblage of permanent buildings and temporary trailers that made up the Livermore campus was a mecca of destructive technology. Teller, who was lab director for two years and served as its de facto leader for decades more, was among the models for the character of Dr. Strangelove in Stanley Kubrick's 1965 atomic farce of that name.

Within the classified inner confines of its security "fence," as scientists called it, scientists exploited the tiny levers of nuclear physics to build some of the most powerfully destructive devices ever made, in the name of national defense. "Outside the fence," as researchers would say, Teller had assembled a cadre of fundamental scientists whose job was ostensibly to understand how unleashing such unimaginable forces, and other human influences on the environment, might affect health or ecological systems. In doing so they ended up laying some of the key groundwork for modern climate science.

Of course, the focus was the application of thermonuclear force, and Livermore joined Los Alamos, in New Mexico, as one of two scientific bulwarks against the Soviet technological threat. Among scientists it was Teller, said *Time* magazine in 1957, who had "worked hardest and most belligerently to send a warning that the Russians were coming." Livermore also served as the flagship in a variety of national missions under Project Plowshare, a federal program to deploy nuclear explosions for peaceful uses. The program envisioned using nukes to cut railroad tunnels and highway passes, stimulate the flow of oil, and blast ditches for canals in Mississippi, Nicaragua, Southeast Asia, and Egypt. "If your mountain is not in the right place, drop us a card," Teller told reporters in June 1959 before he flew to the site of Project Chariot, a controversial effort to create a harbor near Point Hope in northern Alaska. After national outcry from environmental groups, Project Plowshare effectively ground to a halt.

Teller did defeat his activist rivals, though, on November 6, 1971, when Livermore successfully tested a 5-megaton nuclear warhead at the bottom of a 1-mile-deep shaft on the Aleutian island of Amchitka, an exercise the government said was required for national security. The experiment was conducted despite the protests of earthquake and tsunami experts, the Environmental Protection Agency, and four other federal agencies. The opponents had cited risks that the blast would contaminate groundwater, open "fractures and fissures," and release radioactive material at levels "100,000 times the permissible concentration," though this did not happen.

Livermore hummed not just with the violent echoes of atomic blasts but also with the ministrations of scientists conducting some of the most important fundamental studies of the atmosphere in the postwar era. The practical mission of predicting and tracking radio nucleotides in the atmosphere after nuclear explosions was one motivation for the lab, and Teller had an intuitive curiosity about how the climate system functioned as well. (One of his tasks during the Manhattan Project had been to calculate whether a nuclear explosion would ignite the atmosphere.) "Anything he was interested in was fair game for the laboratory to work on," says longtime Livermore hand Cecil "Chuck" Leith, who built one of the world's most sophisticated computer models of the sky in the 1960s, dubbed the Livermore Atmospheric Model. Leith's group went on to make fundamental discoveries on atmospheric modeling, turbulence, computer graphics, and even the atmosphere of Mars. MacCracken, for his part, converted the Livermore simulation into a climate model and used it to develop theories about glacial cycles. "Teller just thought it was important to try to understand what caused ice ages," says MacCracken.

Teller and his fellow weaponeers listened closely to their scientist colleagues, even if their focus was not on the essential mission of the laboratory. Studies of atmospheric transport made clear that depending on the weather, radioactive particles after a nuclear blast could experience what scientists called "rainout" and fall in dense patches close to the blast site. Since nuclear strategists had weapons in place in Europe designed for medium-range exchanges, they realized that using tactical weapons there could contaminate the environment for both sides. That was among the reasons that bomb makers embarked on a program to make bombs that released less radioactivity when they exploded. By the same token, when atmospheric scientist Julius Chang was examining how nuclear tests might have affected the ozone layer, he mistakenly simulated the effect on the ozone layer if a thousand megaton bombs were detonated; the ozone layer was completely destroyed. This was to say, a full-scale nuclear war would kill not only the participants but virtually all life on Earth. "The study reinforced the need to move away from very high yield weapons," wrote MacCracken in a history of Livermore's atmospheric science program.

Despite the technological risks his scientists were uncovering, Teller maintained a steady faith in the power of science and engineering throughout his entire career. In the 1950s he had clashed with Linus Pauling about the dangers of low-level radioactive fallout. As the disillusionment with technology and its effects deepened from the 1970s into the 1980s, no proponent fought the rearguard action in favor of technological solutions to America's problems as passionately as Teller. In 1979, after the Three Mile Island accident, the seventy-one-year-old scientist launched a personal campaign to defend nuclear energy in Washington, D.C., preparing for twenty hours per day, he said, until we worked himself into a heart attack from the strain. ("I was the only victim of Three Mile Island," he wrote in a full-page advertisement he published in the *Wall Street Journal* to counter the "propaganda that Ralph Nader, Jane Fonda and their kind are spewing.") In 1981, parlaying his influence with newly inaugurated president Ronald Reagan into a spot on the White House science council, Teller advocated for a missile defense program his lab dubbed Excalibur—the precursor to Star Wars. The concept, as close to a government science-fiction program if there ever was one, was to ignite and power nuclear blasts in space that would power X-ray lasers and shoot down enemy missiles.

While Livermore's scientists were outspoken in the 1990s over the threat of climate change, Teller got a different message, and never quite accepted the standard global warming mantra. ("We had to keep working on him," admits MacCracken.) Undeterred, Teller combined a selective reading of his labs' findings on atmospheric science with his techno-optimistic bent, and proposed research into the Pinatubo Option in 1997. The program would cost merely $1 billion per year, he and two colleagues wrote, and they cited studies that estimated the cost of mitigating carbon dioxide emissions at a hundred times higher. Made-to-order aerosols could either mitigate the impacts of "greenhouse gases"—the scare quotes are theirs—or, with modification, *enhance* the greenhouse effect to prevent "the onset of the next Ice Age." He followed it up with an op-ed, again in the *Wall Street Journal*, just as the

Clinton administration was beginning to rally domestic support for the Kyoto climate treaty. The idea of hacking the stratosphere was "not a complex" idea, Teller assured readers, and was a "more realistic" way to solve the climate crisis ("a problem that may not exist," he said in the first paragraph) than having governments tamp down emissions. "Let's play to our uniquely American strengths in innovation and technology to offset any global warming by the least costly means possible."

In 2002 Teller expanded the idea further in a paper submitted to the National Academy of Engineering in which he called "active management" of the atmosphere likely the "most overall practical approach" to keeping world temperatures "reasonable." He continued to advocate for the Pinatubo Option or more sophisticated sunlight-blocking tools than droplets of sulfuric acid, going on to suggest launching "metallic screens so diaphanous as to be literally invisible to the human eye." Sending the devices to the point between Earth and the Sun where the gravitational forces balance, known as "L1," would be the "absolute optimum of all means known to us for ensuring long-term climate stability," he wrote in the paper.

The research in 1997 didn't much influence the debate over Kyoto—the U.S. Senate killed any chance of the United States joining the treaty for reasons unrelated to geoengineering—and if anything, the papers Teller wrote hurt the planethacking research cause. They "weren't done that well," says Cicerone of Teller's geoengineering papers, which were not peer-reviewed but published rather as laboratory or scientific meeting reports. To many, their message—that a technical fix might reduce the need to massively overhaul world energy systems—poisoned the well. And so the idea of geoengineering was tainted with the reputation of its messenger—Teller—for a decade or more. "There's no question, having his name associated with the idea is a bad thing," physicist David Keith told me. But the Livermore taint, he said, had not been as serious as some might have feared. "But in the big picture, the basic trade-off is whether a focus on geoengineering would lead in part to less attention on mitigation—and so far I would argue we haven't seen much of that."

Teller's prose, published only a year and a half before his death, betrayed few such sensitivities. Its mood was upbeat, its optimism brimming. "One thus might say, 'Let's just put a sinking-fund of $1.7 B into the bank for use in generating $1 B/year forever, commencing a half-century hence, and proceed with the human race's business as usual. All of the Earth's plants will be more productive for being much better-fed with CO_2 and much less exposed to solar UV radiation, kids can play in the sun without fear, and we'll continue to enjoy today's climate, bluer skies and better sunsets until the next Ice Age commences.'"

Humanity had discovered the ultimate lever. For Teller, it was just a matter of flipping the switch.

THE SCIENTISTS CHARGED WITH PROTECTING THE AUSTRALIAN sugar industry had a major problem on their hands when they arrived in Puerto Rico in 1932 for a conference. Beetle grubs were devouring their country's sugarcane crops. In Puerto Rico, experts bred a natural predator of the beetles, the cane toad, to control them and protect their crops. The cane toad "can be effectively used as a biological control of the gray background beetle," entomologist Raquel Dexter told the conference. "I strongly advocate the effective use of this amphibian immigrant."

Hearing this advice, the Australian Bureau of Sugar Experimental Stations imported 102 of the toads from Hawaii to breed them near Cairns, Australia. In 1935, more than three thousand were released into cane plantations in North Queensland. At least two scientists—including one named Froggatt—warned that the toads might run wild and should not be released. But after a brief moratorium, they were rereleased in 1936.

Thus began a reign of terror that was to gradually spread over decades and a third of the continent. The toads ravaged beehives; devoured a variety of local plants; outcompeted less aggressive Australian native toads; and, notably, mostly failed to kill the beetles they were originally deployed to vanquish. The toads are now found over a range of 1 million square miles, in Queensland and the adjacent state of New South Wales and the Northern Territory. They can ooze or spray venom, and have injured humans who have been attacked by them. People who have killed and tried to eat them have also reportedly gotten sick.

In 2005 an Australian lawmaker named David Tollner suggested that residents use "cricket bats and golf clubs and the like" to kill the toads. Animal rights groups protested, suggesting instead that putting them in the freezer would be a more humane way to execute them. "When you talk about animal rights I think you've got to think about the rights of our native animals as well," Tollner said at the time. "A cane toad can cause a slow death in a crocodile or a goanna or any other animal that eats it . . . we've got to eradicate them by any means possible."

The Sucking-1-Ton Challenge

Humanity emits 30 billion tons of carbon dioxide each year, and over his thirty-five-year career Sam Bose has been responsible for far more than his share. And why not? Industrial managers such as he never had a reason to concern themselves with the stuff. Until 2001, when his facilities were shut down, the plainspoken engineer oversaw a set of industrial plants in northern California that produced alkali powder. All told, the operation emitted into the atmosphere more than a ton of carbon dioxide every thirty seconds. (It takes the average person more than two years to exhale the same amount, and an average coal plant roughly seven seconds.) Since 1984, Bose regularly drove the hour and a half west from his home to Moss Landing, near Monterey, where he oversaw work at the Brick Plant, as the locals called the weedy, rusted factory by the sea.

California industrial magnate Henry Kaiser built the factory's kilns, pipes, and silos in 1942 to remove dissolved carbonate chemicals—what makes water taste hard—from the Pacific seawater. He was deriving magnesium, an important ingredient for explosives during World War II. After the war, engineers such as Bose used essentially the same reaction, which requires kilns fired at more than 1,800°F, to produce magnesium hydroxide, an alkaline powder used to make heat-resistant bricks. In the 1990s, competition from China and the decline of California's heavy industries led to the business's demise. Bose watched his friends lose their jobs one by one as he worked for years to clean up toxic spills at the site and haul off 12,000 tons of steel equipment for scrap. As the last employee at the plant, Bose almost literally turned out the lights on his way out. "I was getting ready to find work in another state, move my family," he says.

Then a marine geologist from Stanford University in jeans and boots named Brent Constantz arrived in early 2008 amid the ramshackle buildings and seagulls. His company, he said, was called Calera, less than one year old, with fewer than ten employees. He told Bose that he wanted to turn the abandoned site into an environmentally sustainable factory to make cement, one of the most carbon-intensive materials to manufacture on Earth. For decades, Kaiser's employees had turned the ocean's hardness, carbonate ions, into an alkaline powder; Constantz wanted to turn it back into carbonate, which he'd use to make cement. It's the same process that marine organisms use to create their shells, he explained, and limestone found in nature is the compressed, mineralized form of carbonate from those shells after they die. He'd obtain the most important ingredient, carbon dioxide, from the power plant across the street, which released millions of tons of it a year. "I was intrigued," Bose told me. "What he was doing seemed quite doable. But nobody had suggested doing it before."

Within two months Bose was hired, the abandoned factory leased, and a Calera-owned Moss Landing Cement Company founded on site. In May 2008 a pilot plant there produced its first batches of cement, using CO_2 from canisters. Then Bose and other Calera employees installed a mini coal burner, about as tall as a basketball hoop. In a device called an absorber they mixed the carbon dioxide with the alkaline solution. They installed equipment that dried the chemical slurries, which looked like mocha shakes, into fluffy powder. "A large engineering company told us we would have to spend $30 million to build it," Constantz told me during an interview in his office, the corner spot in a two-story office building in a business park in Los Gatos, California. "I fired them. We did it in a few months for $2 million." Constantz often says controversial things in the northern California calm befitting his pedigree as a rock-climber-mountain-biker-surfer-U.S.-National-water-polo-team-alum. (He'd have gone to the 1980 Olympics were it not for the boycott. "I was the goalkeeper. I once blocked a shot with my face," he says.) "I've never seen a company move so fast," says Bose.

It's the exhilaration of the carbon rush that propels Constantz's drive. Someday soon—in a year or perhaps a decade as dictated by the politics—government rules in the United States will mean that emitting a ton of carbon dioxide into the atmosphere will cost companies something. (It already does for most industries in Europe.) So if a Stanford geologist can offer a way for an electric company to avoid doing that, he'll make a lot of money. Economists estimate a $1 trillion market beckons to anyone clever enough to grab the world's most wanted molecule from a smokestack or a tailpipe or even thin air.

The climate crisis begins and ends with coal. The planet's more than 2,100 coal plants spew out roughly 41 percent of total world CO_2 emissions from energy use. (The 150 largest of those facilities emit a whopping 10 percent of human CO_2 emissions.) Coal is cheap and abundant, and humans have been burning it to produce energy for more than a century. It's responsible for half the electricity generated in the United States, and it's the climate nightmare that is only beginning: by 2030, forecasts the International Energy Agency, world coal power will double. Renewable energy, natural gas, and nuclear energy will all help provide lower carbon emissions in the years to come, but none will replace coal fast enough to matter. And if coal plants are using roughly the same technology in 2030 as they are now, says the Boston-based

Clean Air Task Force, world CO_2 emissions would be 13 billion tons per year higher. If the world can obtain energy from coal without adding to its carbon problem it will be well on its way to stabilizing carbon concentrations. "If we don't solve the climate problem for coal, we're not going to solve the climate problem," Princeton physicist Robert Williams says.

In these early days of this particular frontier, the rules are simple: it's the race to win the sucking-1-ton challenge. Suck 1 ton of carbon dioxide from the business end of a coal plant for a reasonable cost and you've done it. Engineers know how to suck 1 ton from a coal plant before it's emitted, they know where to put it, and they even have techniques to filter the CO_2 from the sky, as unlikely an endeavor as that seems. But it's all just too expensive right now to make it affordable on an industrial scale. A set of analyses conducted for a workshop at MIT in 2009 put the sucking-1-ton cost as high as $100 for a variety of different coal plants. (As the industry gained experience, the costs would go down, at least for the first plants to control their CO_2 pollution.) That extra cost, if passed on to consumers living in states with predominantly coal-fired power plants, could double the average electric bill.

Experts agree that capturing carbon from power plants is a huge challenge. Some are more skeptical than others on how daunting the numbers look. Projected costs for capturing carbon just keep rising, says research engineer Howard Herzog of MIT. Yet energy expert Armond Cohen of the nonprofit Clean Air Task Force, says the problem "is one of the more tractable challenges" in the global warming crisis. Whether the solution is building new coal plants or retrofitting existing ones, the solutions are all years, if not decades, from being proven economically feasible. Vacuuming CO_2 out of the sky, where it is much more diffuse, is much more expensive, not surprisingly. And whether one obtains the carbon from a coal plant or the atmosphere itself, finding a place to put all the carbon we come up with could be a challenge whose difficulty coal experts are underestimating.

A number of studies have suggested that even if humanity stopped emitting carbon immediately, the global temperature would rise 1.2°C or more, passing the 2°C-limit rise that scientists have set as a warming target. (The planet has already warmed 0.8°C since preindustrial times.) That underscores how important the sucking-1-ton challenge is.

Some scientists feel that the situation is slightly less dire than that . . . but only slightly. In 2009, experts calculated in *Nature* that to have a good chance of preventing the global temperature from rising more than 2°C, humanity should emit no more than the equivalent of 1 trillion tons of carbon dioxide between 2020 and 2050. (Call it the 1-trillion-ton challenge.) In the first six years of the twenty-first century, calculated the researchers, humanity had already emitted 236 billion tons. At the rate we are burning fossil fuel, they calculated, the world's nations have twenty-five years to peak their emissions and then lower them—and even less if, as expected, emission trends continue to accelerate. (The 2.4 billion tons of carbon dioxide emitted each year as a result of burned or clear-cut rainforests only makes our task more difficult.) We have to immediately launch a worldwide program to stop polluting our atmosphere with this surprisingly pernicious trace gas. This is among the biggest challenges of our time.

Whether the carbon polluter you have in mind is a coal plant that provides electricity to 250,000 homes or a Camaro, the basic physical challenge is roughly the same. Carbon dioxide is devilishly stable; for most of human history that fact was a blessing for the planet. Its chemical structure has allowed this crucial molecule to persist and keep our planet warm over the aeons. (If it wasn't for the greenhouse effect, the surface of the planet would be at an average temperature of -2°F.) It gains its stability by virtue of symmetry: desirous of carbon's abundant electrons, the two oxygen atoms in a CO_2 molecule pull tightly on their shared partner like a pair of nurses yanking a patient's legs and arms in opposite directions.

(We should call it OCO, as it's a linear molecule—and therefore as slippery, in terms of chemistry, as a water snake.) Only the nastiest chemicals can react with it. That makes it hard for chemical engineers to grab or filter it, unless they blast it with high temperature and pressure.

Carbon dioxide makes up only 10 to 15 percent of the flue gas coming out of the tail end of a power plant burning coal; this is relatively dilute, so reacting with it requires a lot of energy. Filtering CO_2 out of the sky is even tougher, since grabbing such an unreactive gas, making up less than 0.04 percent of air, is doubly difficult. Scientists know plenty of ways to do it; it just takes a lot of energy, which makes it hard to do cheaply without burning fossil fuels and spewing out more carbon.

So they're trying everything out on the Wild West of the carbon frontier. Constantz, whose Ph.D. thesis examined how corals grow, is teaming with guys who manage chemical plants like Bose to turn CO_2 emitted from power plants into cement. Michael Tractenberg, a neuroscientist in New Jersey, has a company that is using biological membranes to grab carbon dioxide from coal plants' exhaust pipes. Physicists from Columbia University are building machines to suck it right out of the atmosphere. Robert Williams, at Princeton, wants to burn biomass from special fast-growing grasses with carbon-sucking techniques to sequester millions of tons of CO_2 each year.

And despite the crowd of competitors, the sucking-1-ton challenge seems decades from being solved. Since 1997, when the passing of the Kyoto protocol set up the world's first greenhouse gas control regime, there has been next to no progress on effectively keeping greenhouse gases out of the atmosphere. Installing technologies that scrub coal plants of their carbon emissions keeps getting more expensive, says Herzog. Underground sites to store the carbon are unproven and environmentalists might oppose CO_2 storage there. More outlandish options are long-shot bets, and the more scientists learn about the options, the harder it is to see how engineers will be able to provide a technical fix to the problem.

• • •

When scientists say "carbon capture," they usually mean first capturing the CO_2 before polluting it into the atmosphere, and shoving it instead into the ground. The sucking-1-ton cost generally reflects both steps, and both are years from proving to be affordable on the massive scale required.

Transforming the world's coal infrastructure will be a huge challenge, but Cohen, the energy expert, points out that the world's power industries have had massive building phases in the past. Between 1950 and 1970, for example, the United States quadrupled its installed capacity to generate electricity with power plants. From 1960 to 1980, by the same token, the nation laid down 150,000 miles of natural gas pipeline. So transformation of the world's coal power plants would not be completely unprecedented and "can be achieved over the next several decades," he and colleagues wrote in 2009.

Most coal plants, including brand-new facilities in Asia, essentially work by burning pulverized coal in a boiler, which drives steam turbines after heating water—a ninety-year-old technique. (By burning the coal at higher pressures or temperatures using state-of-the-art technology, the process can be more efficient.) Grabbing the CO_2 can be done in several ways. The first is to build new kinds of power plants that work differently. Instead of burning coal to produce carbon dioxide, these plants work by using high heat and pressure to turn coal into hydrogen and other gases, removing the carbon dioxide for underground storage , and then burning the hydrogen to run a turbine. The handful of coal plants set up this way, at three-quarters scale, are demonstration projects in the United States and Europe. The first full-size version of a so-called integrated gasification combined cycle plant will be built in Edwardsport, Indiana, by a power company called Duke Energy, slated for completion in 2012. It has been estimated that the construction cost for its $2.3 billion plant will come out at roughly $3,730 per kilowatt of generating capacity. The company does not plan to store its emissions underground, but if it did, estimates Herzog, that cost would rise to roughly $5,000 per kilowatt. That's equivalent to about $150 per ton of CO_2—"way too expensive," he says.

Since 2007, however, experts in the climate and energy field have realized that retrofitting existing coal plants to suck up carbon might be more important than finding better ways to build them. Depending on how fast we burn coal, the world has one to three centuries of reserves remaining—numbers on such things are notoriously murky—and China has built traditional-style coal plants equal in generating capacity to the entire U.S. fleet. Power plants burning coal are contributing more, not less carbon pollution to the atmosphere, as experts estimate that the amount of power generated by coal plants will skyrocket by 65 percent over 2009 levels by 2020, with most of this growth from China and India. This means that thousands of coal plants will need to be altered—or shut down—if the problem is to be solved.

There are two basic ways in which engineers could alter existing coal plants to grab their carbon dioxide. By replacing the power plant's air supply with pure oxygen in the boiler, scientists can change its exhaust from a 10 percent carbon dioxide stream to an almost pure CO_2 exhaust that is relatively easy to grab. (Oxygen is only 21 percent of air.) But the method requires loads of energy to obtain the pure oxygen, and in doing so cuts the efficiency of the coal plant by 36 percent. The alternative, adding equipment onto the end of an existing coal burning plant to essentially filter its exhaust, is probably even more expensive, both in terms of energy lost and dollars. Participants in a 2009 symposium hosted by MIT estimated that adding carbon suckers to existing plants would cost fifty dollars to seventy dollars per ton of CO_2 grabbed. That's much too expensive for Chinese industries to adopt—a huge problem, since that country and India will be the two biggest carbon dioxide addicts for the next hundred years. The trillion-ton challenge looms menacingly.

Since coal experts haven't made retrofitting existing plants much of a priority until recently, it's hard to know whether the technologies that have gotten the most attention from engineers and developers will end up being the best way to clean up these dirty plants. The standard industry technique is to pass the plant's exhaust through a solvent known as MEA, long used to separate CO_2 from natural gas. It's an effective and well-developed method, but it requires a lot of energy, and many experts think it's a crude process that can never be affordable for thousands of power plants.

On a 2007 tour of the coal-fired Warrior Run power plant in western Maryland coal country, the enormity of the carbon-capture task came into sharp relief. Wearing a helmet, earplugs, and safety glasses, I toured the plant with Larry Cantrell, the plant manager, who spends much of each day studying tables listing the plant's daily emissions. Inside its unheated, cavernous rooms, it's a very loud place, but a surprisingly clean one, with most of the sooty coal dust found where trucks each day deliver it to a large pile where it is ground up before burning. Inside the towering plant a relatively small control room features a handful of computer screens and engineers watching them, as well as manual controls with the obligatory old-fashioned lights and switches.

Part of the revenue Cantrell delivers to the company comes from the CO_2 the plant captures and sells each day to beverage manufacturers; it is shipped from the plant in liquid form by trucks with tanks. "If you had a Coke today, you've probably ingested some of our product," Larry likes to tell guests. A pipe roughly two feet wide brings the exhaust about 440 yards from the main building to the facility where carbon dioxide from the plant is captured. The carbon scrubbers are bulky cylinders several yards tall that are packed with MEA molecules. They continually filter carbon dioxide from the plant's exhaust. After the MEA particles do their thing, they net a total of 89 tons of CO_2 each day for the soda. But that carbon booty, 5 percent of the plant's CO_2 emissions, comes at a price. The Warrior Run plant already produces roughly 10 percent more energy than it can sell just to keep the plant's lights, motors, and other equipment running. The so-called parasitic load rises by another 2 percent for it to capture the CO_2 it puts into soda. That's electricity it can't sell to customers.

Scaling up to capture nearly all the carbon dioxide from big power plants and shove it into the ground would cost much, much more energy. Households would eventually shoulder the added costs. Recent work done by California expert Dale Simbeck puts the sucking-1-ton-cost at seventy-four dollars per ton of CO_2 for existing plants. (That includes the electricity required to heat the MEA columns, run compressors that turn the CO_2 into a liquid, and electricity to make the whole thing work.)

Other methods for grabbing the CO_2 are promising but even more unproven. The trick with designing chemicals to grab CO_2 is that after they bind the elusive molecule, it must be pried off them to obtain a pure stream. So scientists have been looking into alternatives to MEA that snatch carbon dioxide molecules less firmly and therefore require less energy to be pried off it. Scientists working at Notre Dame have discovered that substances known as ionic liquids are particularly adept at dissolving carbon dioxide. (They're weird substances—molten salts that are liquid at room temperature.) Trachtenberg focuses on carbonic anhydrase, a protein found in the brain as well as many other cells. It grabs carbon dioxide and releases it at much lower energy levels than MEA. But so far the promising results have yet to graduate from the testing phase.

We may well have to suck the atmosphere's carbon dioxide flat out of the sky. After all, the root of the problem is that there's too much of this stuff in the air, so the answer jumps out at scientists: make machines that filter the very atmosphere. Operation DustBust Earth.

It's going to be hard enough to reduce the carbon footprint of power plants, but small diverse ones like cars, trucks, and woodstoves make up another 40 percent of world emissions. Leaf blowers offer starkly different challenges than coal plants from a sucking-carbon perspective. It's unfeasible to alter every car or Toro ride-on mower to catch their carbon. So we have to either use less energy to run them, come up with less carbon-intensive fuels to fuel them, or find ways to suck CO_2 right out of the atmosphere to deal with their emissions after the fact.

Air capture, as scientists call it, is a radical and audacious idea. For chemical engineers, getting reactions to work efficiently is about working with chemicals in high concentrations. So it's a little difficult to imagine that taking carbon dioxide molecules out of the air, where their concentration is one particle in twenty-five hundred, could ever compete with grabbing carbon dioxide at a coal plant, where the concentration is more like one in seven.

Still, there are several reasons why it makes sense at least to investigate. Air capture deals with the carbon spewed out of the tailpipes of cars, which can't realistically be captured on a vehicle-by-vehicle basis. If artificial trees could ever be built, they could be installed far away from cities, and built en masse where construction costs or raw materials are particularly cheap.

Creating a program to draw down the carbon dioxide in the atmosphere would be a massive international undertaking with few historical antecedents. The vision is certainly one of grandeur; it's up for grabs as to how delusional it might be. *Nature* imagined establishing a program to account for half the carbon emissions after 2020, sucking a total of 650 billion tons of carbon by the twenty-second century. The magazine envisioned each sucking station pulling in 250,000 tons of carbon a year, roughly a quarter the size of storage projects in Algeria and the North Sea. To get all that carbon out of the sky would require 35,000 such stations. "A major air capture program would be the biggest public-works project the world has ever seen," University of Michigan law professor Edward Parson says. And if the sucking-1-ton cost for Operation DustBust Earth is closer to the thousands of dollars than the hundreds, it won't happen whether or not it's logistically possible. Even two hundred might be too much.

But during the course of a century, that cost could fall and make the weird idea almost feasible. A thought experiment published in 2009 by University of Colorado policy expert Roger Pielke Jr. illustrated how the cost, spread over the decades, could compare favorably with other methods. His approach was to pretend that society did nothing to tackle climate change, allowing the concentration of carbon dioxide to rise until it reached 450 parts per million—and then turning on an air capture regime full bore. By avoiding the hard steps now—whether by choice or not—he figured, the world economy could grow unfettered by carbon regulations until 2050, when a phalanx of air-capture devices—what he calls the "brute force backstop" of climate policy—could be deployed. For argument's sake, Pielke accepts at face value the sucking-1-ton cost that air-capture advocates have come up with, between $100 and $130. Pielke isn't suggesting that humanity avoid tackling the problem right now—on the contrary, he wants, posthaste, all the solar panels, windmills, nuclear power plants, and reformed coal plants that the world can muster. His point, he says, is just that aggressively trying to lower emissions may not prove sufficient—so studying air capture in case we need it makes intuitive sense.

And so a few groups are doing so. University of Calgary physicist David Keith prefers the "Russian tractor" approach, as he puts it—simple, well understood. His technique is the most chemically obvious, and his goal, he says, is simply to find out how much the whole approach costs. Chemical engineers have known for more than a century that a caustic alkaline chemical called aqueous sodium hydroxide (commonly called lye) will react with carbon dioxide in the air easily to make carbonates—companies started doing it in the 1950s. Keith's process involves a series of similar chemical steps, including one involving a 900°C kiln that essentially pries off the CO_2 from the carbonate. His university team built a prototype in 2005, and he started a company in 2009 called Carbon Engineering to further develop the technique. The goal is to scale up to build plants that could sequester a million tons of CO_2 each year.

As head of his company, he declines to talk about what sucking-1-ton cost he thinks he can achieve. Before he started the firm, however, Keith estimated that four years of research and development could bring the cost below $200 per ton of CO_2. "If we did it under a hundred dollars, I'd be pretty excited," he added.

But Herzog of MIT says that Keith's calculations are "not realistic" and that there's no getting around the fact that the carbon in the atmosphere is just too damn dilute. He puts the sucking-1-ton cost for air capture at $1,000 "or more," he says. Methods for grabbing CO_2 from coal plants, like the MEA solvent, use a so-called absorption column to allow the plant's exhaust to interact with the solvent. Given the difference in concentrations, says Herzog, if a power plant needed an absorption column of a certain size then one would need "three hundred times more cross-sectional area" to grab the more diffuse CO_2 from the sky. In 2006, chemical engineer Marco Mazzotti of ETH in Zürich, Switzerland, calculated that using such a technique would require 12 gigajoules of energy per ton of CO_2 sequestered; burning the coal to produce that 1 ton of CO_2, however, gives off only 9 gigajoules. That makes this approach at best "a hypothetical long-term technology," and a flat-out energy loser at worst, says Mazzotti.

Other teams have gone for more radical approaches. Swiss scientists are trying to use the energy of the Sun to grab carbon, utilizing curved mirrors to heat air capture reactors to 800°C. Columbia University physicist Klaus Lackner uses modified sheets of a commercially available filter surface that binds CO_2. In experiments in a warehouse outside Tucson, Arizona, in 2007, Lackner discovered that the surface grabbed CO_2 in dry air but then tended to release it when the humidity increased. By harnessing the energy of evaporation, he says, he doesn't need much additional energy—perhaps only the power of a fan—to draw air through a box in which the surface sits.

Lackner has proposed collecting CO_2 on the surface in dry outdoors air and inserting it into the moist environment of a greenhouse, where the CO_2 would be released and would be taken up by plants. Growers pay more than $100 per ton for concentrated CO_2, and Lackner hopes that providing the gas for greenhouses will provide his company with a niche market to eventually scale up. Michael Trachtenberg, a competitor with Carbozyme, in Monmouth Junction, New Jersey, told me it's "a very clever update" of older techniques.

"Nature has already designed perfectly good air-capture machines: plants," MIT postdoctoral scientist Kurt House told me. Lackner knows that by fertilizing greenhouses with the CO_2 he sucks out of the air, he isn't storing it permanently at all—he's just sending it through the biological cycle one more time. Eventually the plants from the greenhouse will decompose and release the carbon dioxide he's fed them back into the air.

The most effective way to deploy plants to solve the sucking-1-ton challenge is to grow them, burn them for energy, and sequester the carbon dioxide produced deep in the ground. Like the rest of the approaches, turning biomass into carbon-negative fuels or electricity has never been tried on a commercial scale. But it would definitely work. As long as scientists capture the carbon dioxide they produce when they roast the plants, blasting biomass under high pressure and heat is the best way to utilize their prodigious carbon-sucking abilities. The process is known as gasification—it's very similar to gasifying coal—and it involves turning biomass into "syngas," a mixture of hydrogen gas and carbon monoxide molecules. The sucking-1-ton numbers that scientists have estimated are pretty optimistic: an American company called Rentech, for example, says it can turn a mixture of coal and biomass into carbon dioxide for six dollars a ton, provided it can sell the carbon dioxide to oil companies who use the gas to tap old oil wells and get extra petroleum out of it. That solution might help the industry get going, but scientists say that to store all of the coming carbon emissions from coal, society is going to need a lot more space than is available in low-yielding and abandoned oil fields.

But a tainted past haunts the promising technique of gasification, and the method could have an ominous future. An ill-conceived effort by the U.S. government in the 1970s to turn coal into "synfuels" as an alternative to gasoline became a symbol of government boondoggles after Congress shut it down in 1986. Worse, using the method to gasify coal to make liquid fuels can be lucrative when the price of gasoline is high, but it results in a disastrously high carbon footprint. Still, the Chinese government, the U.S. Air Force, and a number of firms are pursuing making synthetic gasoline or other fuels out of gasified coal.

From a climate perspective it's an atrocious idea, since it leads to emissions of CO_2 in the production of the fuel as well as in burning it. Supporters of the technology, who have lobbied Congress for tax incentives and rebates, say that making fuels could lead to a sucking-1-ton price of less than forty dollars. But environmentalists such as David Hawkins of the Natural Resources Defense Council feel that encouraging companies to build gasification plants with the promise that they will eventually gasify biomass is foolish because the firms could use coal to make fuels. That leads to roughly twice the amount of carbon emissions that burning coal would have without making it into liquid fuel.

Whether engineers end up capturing carbon dioxide from the rear end of a power plant or right out of the sky, they'll have to put it somewhere. That's supposedly the easy part, but as competitors try to attack the sucking-1-ton-challenge, it's becoming clear that storing underground the carbon dioxide engineers managed to capture could be a significant headache.

There's the sheer logistical enormity of the challenge, for one thing. Creating a system to store U.S. carbon dioxide pollution from power plants alone would be a massive public undertaking, akin in scope to, say, setting up new railroads all over the country. Shoving carbon dioxide from the air into the ground would require an even more massive endeavor. Collecting 60 percent of the carbon dioxide produced from American coal plants and preparing it for underground storage by liquefying it would produce roughly the same volume of liquid as total U.S. oil consumption: 20 million barrels a day. (Optimists point out that engineers store three times the *total* volume of CO_2 emissions in briny liquids from oil fields each year—so it's doable, they say.) No wonder the U.S. petroleum industry is so ambivalent about coming greenhouse gas regulations: firms such as Halliburton, with their expertise in exploring Earth's crust, will play a huge role in the next chapter of the U.S. energy story.

Since roughly 2005, scientists have focused on putting carbon dioxide under pressure and squirting it in liquid form into rocks hundreds or thousands of feet underground. The target would be porous rocks that sit below impermeable layers. Limestone and other carbonate rocks have tiny holes in them. Fluid, usually brine, often flows through them, and if you own the land and you're lucky, black gold. For decades oil companies have injected carbon dioxide thousands of feet underground to drive petroleum to the surface. But they've injected that CO_2 in small amounts, sporadically, and geologists have rarely examined the fate of the carbon dioxide they've shoved down. As of 2009, scientists had attempted large-scale CO_2 storage demonstrations at only three sites around the world. Demonstration projects in Canada, Algeria, and Norway, which obtain their carbon dioxide not from coal plants but from various commercial efforts to collect fossil fuels, have each injected roughly 1 million tons of carbon dioxide into the surface each year, with few reports of leaks or other problems. To put this number into context, consider that coal plants around the world produce more than 9 *billion* tons of CO_2 a year.

Injecting millions of tons of liquid carbon dioxide deep into Earth might bring some risks with it, but scientists can't yet put numbers on them. Injected deep enough, there appear to be few threats to groundwater, says geologist Susan Hovorka. "God knows, we've looked for problems," she says. Decades of injecting millions of tons of carbon dioxide into wells in Texas have never resulted in an accident, though the possibility of a large release of carbon dioxide could, at least theoretically, asphyxiate nearby residents. (A huge bubble of naturally produced carbon dioxide rose from

Lake Nyos in Cameroon in 1986 and killed seventeen hundred people, though geologists say this is not a risk with carbon storage.) Furthermore, a 2007 report by MIT found that "risks appear small" of groundwater contamination, though "the state of science today cannot provide quantitative estimates of their likelihood." There's also the real possibility that injecting millions of tons of fluid underground might cause earthquakes, but rules that the Environmental Protection Agency already has in place require careful analysis and monitoring of storage areas.

While these dangers seem remote at best—especially compared to the very real dangers that mounting CO_2 levels pose—their very existence has fed a small but potentially significant movement of activists opposed to sequestering carbon underground. Vattenfall, a Swedish power firm, had planned to inject 100,000 tons of carbon dioxide captured from a coal plant by April 2009 into the ground below Spremberg in northern Germany. The $100 million project, known as the Schwarze Pumpe, would have been the first effort in the world to capture carbon pollution from a coal plant and inject it directly into the ground. But delays from local safety boards, lobbied by citizen groups, have pushed the project back a year, and engineers on-site are venting the carbon dioxide into the atmosphere instead of the ground. As of July 2009, a Vattenfall spokesman didn't expect a permit for injection until spring 2010. "This is a result of the local public having questions about the safety of the project," he told the British *Guardian* newspaper. "People are very very skeptical." (The paper called it NUMBYism: Not Under My Backyard.) The protests opposing the German project, said an activist with Greenpeace, were "a stark warning to those that think CCS [carbon capture and underground storage] is an easy solution to the huge climate problems of coal-fired power stations."

The setback followed a similar development in the Netherlands, where a local council nixed a similar project. Dozens of small- and medium-scale projects have gotten underway successfully in the United States. But perhaps foreshadowing a larger debate on carbon storage in the United States, a vocal group in rural western

Ohio near Greenville in 2009 opposed an underground carbon storage demonstration project planned there. "What's the rush, the public deserves to be safe," wrote commenter Michael William on a blog organized by opponents. In August of that year, under political pressure, the company proposing the $93-million project abandoned it, citing "business considerations."

On the outskirts of Silicon Valley, Brent Constantz is scraping for turf on the carbon frontier and taking his own potshots at underground carbon storage. "Injecting massive amounts of CO_2 underground is Russian roulette," he declares over a pizza dinner with me and a few of his employees one August evening, his cement company less than two years old. A week before, Constantz attacked geologic sequestration in congressional testimony, asking lawmakers to "level the playing field" and provide the same federal dollars for cement-making as they're spending on traditional modes of disposal.

When Constantz began his company, ironically, it wasn't storing carbon from big polluters that he had in mind. For decades he had dreamed of using the carbonates dissolved in the water of the ocean to make rock, the same way marine organisms use it to create their shells. "In the eighties I was thinking about building overpasses in the Red Sea—whole structures, out of pearl, right out of seawater," he told me.

His original idea for Calera was simply to make cement that required less carbon emissions than the standard mix. (The global cement industry emits 1.1 billion tons of carbon dioxide each year.) He first pitched the idea of Calera to renowned venture capitalist Vinod Khosla in a 144-word e-mail on a Saturday morning in June 2007. "I have an idea for a new sustainable cement that would replace standard Portland cement," he wrote. The cement he wanted to sell "would remove a ton of carbon dioxide from the environment for every ton of cement produced." Constantz received one of the autoreplies for which Khosla is well known in

Silicon Valley: sporty and zen, he likes to tell people he's on vacation. But the next day, Khosla wrote back. "I would love to receive any materials you have on this," he said. "Green cement is high on my priority list." Within two months Khosla had become Calera's main investor, and two months after that Constantz had assembled a team of ten people.

In the fall of 2007, Khosla was visiting the company's fledgling facilities, his standard black turtleneck peeking out from the white lab coat the Calera employees had provided him. At the time, the game plan was simply to lower the carbon footprint of a ton of cement. Constantz was showing his investor various mixtures of cement the team had created. The team had been trying to combine calcium and carbonate, two ions found in seawater, to make a solid. But they weren't having much luck—except for an experiment in one bottle, into which the scientists had bubbled lots of carbon dioxide. Constantz asked a cement scientist, Laurence Clodic, how that sample had performed. "Eight times the yield," she reported. Constantz looked at Khosla. "That's great, Vinod. Where can we get a lot of carbon dioxide?" It didn't take them very long to realize what the experiment suggested. "This was a green cement company," says Constantz. "All of a sudden we were a carbon sequestration company." To make concrete, one mixes water with cement, essentially the glue, and adds filler in the form of sand or gravel. Within seven months, Constantz employees had mixed up the first batch of Calera concrete, poured into a tube roughly six inches thick. Slices of the concrete tube exhibit the Calera cement's gray color with filler rocks of red, blue, and green. In the first two years of its existence, Calera has grown from twelve to seventy employees, including eighteen Ph.D.s and five patent specialists. Khosla has poured more than $50 million into the company.

The firm is spending as much as $2 million a month, it says. Is the fact that the company's path has already shifted directions a number of times an indication of admirable flexibility or lack of direction? Going from the cement company to a sequestration company was only the first tack. In the beginning, Calera aimed to use seawater to drive the process; before the firm was a year old, Constantz had decided that seawater wasn't crucial. Then the team realized they could make not only limestone for the cement but also use it as the filler. "The amount of paradigm shifts that have gone on every few weeks here is crazy," says materials engineer Chris Camire, a cement expert with the company; his face lights up when the conversation turns to the Constantz. "He's like a chess player: he sees five, six, eight moves into the future." The company has notched several breakthroughs as it has shifted course, says Constantz, including designing a particularly effective absorption device the team calls the "KazBob," after its inventors, whose names are Kaz and Bob, respectively. (Its tubes give it a somewhat caterpillar-like resemblance. "It looks like some kind of medieval thing," says Camire.)

The challenge for Constantz will be proving to skeptics that the process can be profitable at the massive scale he imagines. ("I hope he becomes a billionaire," says Stanford climate scientist Steve Schneider. "That would be a very good thing for the planet.") Constantz had his first public spat as Calera chief when he sponsored an exhibit at the California Academy of Sciences in which he described the inputs to the Calera process as carbon dioxide from power plants and seawater. What was missing, pointed out Ken Caldeira, was any mention of alkalinity, required to make the process work without causing the ocean to emit carbon dioxide. "Calera and the Academy of Sciences are now misinforming schoolchildren," he wrote on an online message board. "When I raised these concerns to Calera, they would not respond openly to my critique, asking instead to sign a nondisclosure agreement." Constantz dashed off an angry response, rife with spelling errors, on his iPhone. He questioned Caldera's "personal imtegrity" and mentioned a patent Caldera had applied for 2001 on a related technology. More details on his firm's technique, he said, were proprietary. Constantz accused the climate scientist of pressuring his firm and disguising "a greedy hope of a royalty stream as a concern for schoolchildren."

But the company knows it has an alkalinity challenge, to say the least. Wrenching the ultrastable carbon dioxide molecule into carbonate takes ultra-strong sour solutions. One of the main reasons why Constantz set up shop at Moss Landing is that a few hundred feet from Calera's pilot facilities are giant white meadows of alkaline powder, industrial waste known to the locals as Moss Mag.

Calera's trickiest problem will be coming up with the stuff elsewhere. "Not everyone has 5 million tons of Moss Mag sitting around to react with CO_2," Constantz admits. Other sources of the base it needs include the ash by-products of coal-burning and cement-making, but there's not enough of either to make a global impact on the problem of carbon emissions from coal. To get alkaline solutions manually, the company must zap water with electricity to produce the ions it needs, a process known as electrochemistry. For this purpose Constantz says his company is producing applications for "two patents a week" in that area, the fastest-growing part of his firm. Several top electrochemists sit on his advisory board. "A year ago we weren't even thinking about e-chem," he says.

One of their breakthroughs is a clever way to combine two chemical reactions to produce carbonate ions more efficiently than they have been made in the past. It's the brainchild of a twenty-two-year-old wunderkind who often wears a backward baseball cap and ripped jeans. The company plans to test the new technique in a $10 million pilot facility in 2010 at the Moss Landing site.

And the sucking-1-ton-challenge? Constantz claims he can achieve the almost mind-blowing cost of seventeen dollars using the bicarbonate method. If he can do it in a few power plants, he'll be considered a major success for a start-up company. If he can do it in many power plants, he could become a household name. And everywhere one looked would be a building or a road made from carbon pollution. To save the world, we would turn carbon into carbonates, making limestone as a component of concrete, or injecting a solution of bicarbonate into the ground like a big mocha shake. "This isn't just a niche solution," he tells me in an interview. "We will be the primary solution."

A number of experts say his numbers are bogus. Calera's ability to become the answer to the world's carbon glut will depend heavily on obtaining alkalinity. In places where there aren't mountains of Moss Mag or other low-cost alkaline chemicals sitting around, making the alkalinity it needs to react with carbon dioxide will require running electrochemistry facilities next to power plants, sapping their energy. Instead of making concrete, they'd be making the bicarbonate solution—the mocha shake—which would need to be injected underground, which could be more expensive than making building materials that they could presumably sell. To get that magically low seventeen dollars per ton of CO_2 sequestered, the company would be monopolizing a quarter of a coal plant's electrical output. Calera says it could use a power plant's electricity at night, when it's cheap.

But doing electrochemistry to obtain the alkaline solutions Calera needs means creating large amounts of hydrochloric acid, one of the nastiest industrial wastes around. Also, Harvard geochemist Dan Schrag estimates that injecting a ton of pure CO_2 into the ground delivers more than twenty-five times the amount of gas than injecting a ton of bicarbonate, which is only roughly 4 percent CO_2. (Calera says that pumping bicarbonate solution as a partial solid can make up the difference—but geologists worry that injecting solids into the ground will seal up pores in the subsurface rock layers.) And there's the real possibility, say geologists, that the bicarbonate will react with briny water underground, and out will bubble the carbon dioxide. "If something sounds too good to be true, it probably is," says Howard Herzog. As for the seventeen-dollar sucking-1-ton cost? "My initial reaction is he's pulling numbers out of his ass," says Schrag, who has met with Constantz and actually licensed a patent to him.

Constantz knows that mainstream climate scientists such as Schrag are skeptical of his work, but he says that in general, the field of climate science needs more outsiders like him because it's grown stale. "In terms of good-quality, high-impact science there's barely any of it," he says, sipping a beer from a tall mug. On his iPhone he flips through vacation photos from Lake Como, Italy, where limestone is everywhere. Limestone walls. Limestone steps. Limestone patios. You can see that in his mind, the geologist is imagining that the steps out from the pizzeria we're sitting in will one day be built of pure limestone, made from the carbon emissions of a powerplant that provides the electricity that keeps the pizza ovens hot.

MORE THAN A HUNDRED BOATS CONVENED ON A SPRING DAY in 1972 to dump worn-out tires off the coast of Fort Lauderdale, Florida. Marine biologists and engineers had planned to grow a man-made reef to provide a habitat for fish. "Tires, which were an aesthetic pollutant ashore, could be recycled, so to speak, to build a fishing reef at sea," said Gregory McIntosh of Broward Artificial Reef Inc., explaining the concept. A Goodyear blimp dropped a gold-painted tire to signal the beginning of the effort, which was enthusiastically publicized by the tire company. Over several years of dumping tires, roughly seven hundred thousand were strewn about the seafloor across seventeen acres of the ocean.

Fish never inhabited the tires; they moved around too much. Their bindings, which connected the tires to one another, broke, and the tires rolled with underwater tides, crashing into a natural reef near shore, killing mostly everything in their path. The underwater dump spread to cover an area of thirty-five acres. Hurricanes in 1995, 1996, and 1998 drove thousands of tires ashore onto beaches. The effort to clean up the mess, which has involved the U.S. military, inmates, state workers, and shrimp boaters, has cost the state and federal governments millions of dollars.

"It was one of those ideas that seemed good at the time," said Jack Sobel, a scientist with the Ocean Conservancy in Washington, D.C. "I think it's pretty clear it was a bad idea."

A variety of other artificial reefs around the world have been created using everything from sewer pipes to ditched airplanes and boats to the sunk USS *Oriskany*, an 880-foot aircraft carrier dumped into the Gulf of Mexico in 2006. "There's little evidence that artificial reefs have a net benefit," says Sobel.

Credit Is Due

It was to be the launch of a "Voyage of Recovery," an "unprecedented ocean science effort to slow global warming," the press release announced. Reporters were invited; the National Press Club in Washington, D.C., was to be the venue. It was March 12, 2007. Russ George, geoengineer and entrepreneur provocateur extraordinaire, stood at the front of a small room as the event began. "Planktos Eco-Restoration," a slide read. A smattering of reporters listened. Another slide depicted the 115-foot *Weatherbird II*, the company's research vessel, with the company's name, Planktos, emblazoned in large white letters on the black hull. George, a Canadian environmentalist with a goatee, sought to grow massive blooms of algae by adding minute levels of iron nutrient to the ocean. As they grew, catalyzed by the iron, the algae would bolster marine ecosystems while sucking carbon dioxide from the atmosphere. The site of the experiment was to be three hundred miles off the Galapagos Islands, in the tropical Pacific.

George explained to reporters that the "product" he was selling was carbon credits for the algae he would grow. A little iron sprinkled in the ocean could go a long way, he said. "We will be producing millions of tons of this valuable product." Each of those tons could be sold as carbon credits, perhaps someday on the international market under the Kyoto protocol. "You actually can save the world, and make a little money on the side in this business. In fact, that's our corporate mantra." George was a for-profit member of the Blue Team.

Given the controversial nature of the business that George was proposing, reporters expected him to line up scientists who would indicate their support for the idea, as is customary in the Washington influence and public relations tradition. But the speakers Planktos had assembled for the event were not scientists, and they did not seem to know much about the company. Noel Brown, a diplomat who had formerly headed the UN Environmental Program, was the featured speaker. He wore a dark jacket with a light pocket square. He had distinguished gray hair and boxy glasses. He was there to "listen and to learn" about Planktos, though he called the company's plan "a most exciting venture" he was proud to support. A distinguished-looking John Englander, head of the International Seakeepers Society, presented George with a certificate of "Founding Membership" in his charity, an honor that the Society's Web site said required "a one-time contribution of $50,000." The Galapagos project, to be carried out over four months over a 3,900-square-mile swath of ocean, was to be the first of six giant experiments George envisioned. Others would target different areas of the world's seas. In each experiment, Planktos crew members would grow the algae by dumping into the ocean more than 45 tons of fine iron ore, dissolved in seawater. The "marine forests," as George called them, would be so huge as to be visible by satellite, and like plants on land, they would utilize carbon from the air to form their cells. Theoretically, he said, a portion of the bloom would die or get eaten by other microorganisms and fall to the deep as carbon-rich snot and fecal pellets, sequestering the carbon. "It's going to be a world saved by plankton poop, or something like that?" asked a reporter. "Marine snow," corrected the Planktos press aide. Taking carbon out of the sky where it was a menace and allowing it to feed ocean ecosystems amounted to "eco-judo," said George proudly.

Greening the ocean to collect carbon—or cash—sounded improbable, but Planktos had sent to prospective investors a three-ring binder full of scientific papers that explained the origin of the idea. And what investors read was that the concept of fertilizing the ocean had shed light on one of the biggest mysteries of twentieth-century oceanography. The enigma was why whole swaths of the ocean were enriched with bountiful supplies of nutrients such as nitrogen or phosphorus, but little marine life was found there. Most conspicuously, the areas featured few large blooms of algae, the foundation of marine food webs. California oceanographer John Martin hypothesized in the late 1980s that the answer was a shortage of iron, supplied on Earth from dust. Groundbreaking experiments proved Martin's hypothesis and suggested, unexpectedly, that the iron in dust from land might control marine ecology and by extension play an outsized role in global climate cycles.

Budding geoengineers such as George didn't revere Martin simply because his discovery had rendered oceanography textbooks out of date. The iron hypothesis suggested a natural lever that might allow humans to control the climate system. Martin had dubbed it "the Geritol solution" to global warming, a reference to the popular iron supplement sold to senior citizens. "Give me half a tanker of iron, and I will give you an ice age," he half-joked in 1988 ("in my best Dr. Strangelove accent," he said later). In a 1993 experiment that Martin designed to test whether iron could grow algae blooms, scientists mixed a ton of iron dissolved in seawater into the warm waters of the equatorial Pacific Ocean. Days later, recalls Martin's colleague Ken Johnson, their twenty-three-person research vessel was surrounded by green-tinted water. "This is not the ocean we came to," Johnson recalls thinking, with a measure of shame at the time. "It even smelled different." The oceanographers had conducted the test 250 miles southwest of the Galapagos Islands because it featured high amounts of nutrients in the water but little marine life. (Russ George had targeted the area for the same reason.) Over the next decade, international teams would go on to conduct a dozen small-scale iron experiments along similar lines in various seas to better understand the role the trace nutrient played in oceans.

If it reaches a depth of roughly four thousand feet, falling dead algae or waste from its predators might store the carbon for dozens or even hundreds of years, theoretically. Water sitting below that depth will stay unexposed to the atmosphere for that long, so the carbon in the water won't have a chance to get added to the atmospheric carbon cycle until it rises. Some scientists believe that iron fertilization, deployed on a grand scale, could grow vast plankton blooms and pull in a whopping 10 percent to 20 percent of the world's yearly carbon pollution. George's bottom line? He was selling carbon credits for an absurdly low sucking-1-ton cost of only $5. "A typical family of four in North America emits about 20 tons of carbon dioxide in one year," he said. "If your family of four has a 20-ton footprint, that's one hundred dollars per year."

A reporter asked whether any international rules prevented their project. "The concentrations of the materials we put in our almost homeopathic medicine," George answered. The last question was whether George had any business competition. "We think that there are a couple of other companies working in secret," he said "We hold press conferences and are very public about our work. It invites people to come. We live in the world of the high seas. On the high seas there are desperadoes out there, and we're sure that we'll be victims of them sooner or later." That afternoon a flock of media and curious politicos had a short tour of the ship. Three days later the crew was in Norfolk, Virginia, getting ready for their big journey, which was to begin two months later.

Providing an incentive for companies to clean up their carbon act has been tricky enough in the early days of the world carbon market. Determining whether companies should profit off geoengineering schemes such as hacking ocean food webs will be harder still.

When it comes to coal plants and other big carbon emitters, the sucking-1-ton cost, as well as the rules of the carbon market, determine whether companies clean up their carbon pollution. That's how countries are tackling emissions under Kyoto, and presumably under the system that follows.

What Russ George was trying to sell was another kind of emissions credit in which companies and individuals trade: offsets. Offsets are carbon credits that a company can use to achieve its emissions goals that signify reductions achieved through special projects, generally in the developing world. The idea is that a cement plant holding emissions credits for 500,000 tons of greenhouse gas might find one year that it is on track to emit 600,000 tons. If it doesn't deal with the excess pollution, it might have to pay heavy fines, so the plant must buy 100,000 credits one way or the other. It could buy them on the open emissions reductions market, or it could purchase offsets on the international market. In Europe in 2008, for example, those credits were worth roughly $25 per ton, so the cement plant might have spent the $2.5 million to buy credits there. The credits may have represented a variety of Third World projects, including solar power installations, or tree-growing efforts. Essentially they represent emissions cuts that occur in the developing world instead of in the developed world.

Theoretically, a standard emissions credit helps lower greenhouse gas emissions because its price can directly determine how much greenhouse pollution a company has prevented from escaping into the atmosphere. If the emissions cap for a particular coal plant is 500,000 tons, purchasing a credit for any additional greenhouse pollution means that someone, somewhere, is cleaning up their act, at least on paper. Offsets have a murkier value, and face more questions about their legitimacy. Instead of buying new emissions controls for their coal plant so as to avoid buying expensive credits, a company might instead buy offsets that involve, say, growing trees in China. Both actions—cleaning up the power plant and growing trees—might result in fewer tons of carbon dioxide in the atmosphere. The problem is that the trees might have gotten planted anyway, or they might get chopped down if they're not watched closely, or by growing the trees in China, a lumber company will chop down others, in Russia. To guard against these possibilities, official offsets are each certified by an international body and verified by a third-party firm, who certify and monitor the projects.

Offset credits have been among the most controversial elements of the Kyoto protocol, which went into force in 2005. According to the World Bank, Kyoto offsets have produced roughly $106 billion in clean energy investment in developing countries through 2008. That's five times as much investment in that area as more traditional aid sources. In principle, that total shows that offsets might allow fully developed nations to subsidize the development of poorer nations in a sustainable way. The best kind of offsets directly support alternative energy infrastructures such as wind projects or solar power facilities—projects that might have a lasting influence on the way developing countries grow their economies. But billions of dollars in Kyoto offsets have gone to projects with questionable lasting climate benefits. An estimated 17 percent of Kyoto offsets by 2012, for example, are projects using technology to destroy greenhouse gases called hydrofluorocarbons, which generated huge profits for a number of Chinese companies that switched equipment in their refrigeration plants to do so. "It is important to prevent these gases from being vented," said environmental group Friends of the Earth in a 2009 report. But using Kyoto offsets to do so "does nothing to move developing country infrastructure away from a high carbon path." Experts who support offsets say that the hydrofluorocarbon offsets have run out and more long-lasting projects are all that remain. But Friends of the Earth says that other projects, such as hydropower installations in China, "would have happened anyway" because of government energy policies.

If they were ever shown to work—and it's a big if—iron fertilization projects could be the most lucrative climate offset projects of all time. This is why other companies had tried to get rich with the method before George. In 1998, a company called Ocean Farming had conducted successive iron fertilization experiments in a pair of roughly 3-square-mile patches in the Gulf of Mexico, but failed to grow enough algae to boost fish stocks, its goal. By 2001, three start-up firms—GreenSea Venture, Carboncorp USA, and Ocean Carbon Sciences—had explored use of the technique to sell carbon credits, though none ever conducted large-scale operations.

What makes iron fertilization so irresistible to entrepreneurs is the ecological leverage inherent in Martin's figurative one tanker of iron. The president of Ocean Farming wrote in 1998, for example, that fertilizing a patch of ocean off the Republic of the Marshall Islands in the western Pacific Ocean would sequester "30 percent of the CO_2 produced by the United States from the burning of fossil fuels." Scientists say that number is almost certainly too optimistic. But a decade later, when Planktos was introducing itself to the world, scientists had subsequently shown that they can grow big blooms of algae, but not, crucially, that they could suck much carbon permanently out of the sky with them. Four scientists wrote in *Nature* in 2009 that growing a man-made bloom would create a change at the base of the food web that would propagate throughout the ocean ecosystem in unpredictable ways. Moreover, nutrients such as nitrogen and phosphorus would sink along with the carbon, altering biogeochemical and ecological relationships throughout the system. Some models predict that ocean fertilization on a global scale would result in large regions of the ocean being starved of oxygen.

Since scientists do not agree on how harmful the side effects of the technique would be, it's difficult to determine whether the climate benefits would be worth it. And quantifying those benefits is impossible right now, as it's unclear that scientists will ever be able to sufficiently quantify the amount of carbon sucked in by algae blooms to do so. Or, like the Chinese hydropower, that natural algae blooms wouldn't "have happened anyway" if natural bits of iron were delivered to the area, as a component of dust from icebergs or wind. A large-scale experiment, MIT ecologist Sallie Chisholm and colleagues wrote in 2001, would be plagued by inherent uncertainty, meaning that the technique should never be eligible for carbon credits. (Iron fertilization has attracted quite the Red Team contingent.)

Whether or not they thought that Russ George was the right individual to serve as the public face of the controversial field, other scientists saw the possible effects of iron fertilization as less dire. Large areas of the ocean are depleted of oxygen naturally, they point out, and blooms of algae have formed in the ocean for billions of years, in large part because of iron naturally delivered by wind. "People have this pristine view of the ocean," said Ken Caldeira. "They do things on land they would never think of doing in the ocean. If you build a building you kill a lot of earthworms."

The Anthropocene will be full of budding geoengineers such as Russ George, persuasive to some and reckless to others. Changing the globe's climate deliberately, should humanity someday decide to do so, may be impossible without entrepreneurial planethackers.

Planktos, the Greek word for plankton, also means "to wander." It's a word that fits Russ George's career. He lived on a houseboat in Half Moon Bay, California, when he formed his company. He filmed environmental documentaries and volunteered with Greenpeace in the early 1980s. Later, he'd made a living peddling various environment-and-energy businesses, both mainstream (reforestation projects in Canada) and fringey (cold fusion in New Mexico). His next obsession was iron fertilization. On a June day in 2002 he assembled a few Greenpeace buddies aboard the *Ragland*, a 100-foot schooner loaned to the group by Neil Young. (George cited the singer's 1979 album *Rust Never Sleeps* in a press release.) The crew set sail for Hawaii. Along the way they'd pour into the ocean 500 pounds of fine iron dust that George had purchased from the Hoover Paint Company; the dust was sold as pigment. The experiment left a meandering red trail in the wake of the turn-of-the-century sailboat. George called the operation "Planktos IronEx 1," and claimed the following year that it had made the town of Half Moon Bay carbon-neutral for a year. Researchers pointed out that he hadn't monitored the patch of ocean after the

Ragland operation to see whether the experiment had grown algae—a detail required to know whether the process could potentially sequester carbon. ("Someone I know said they saw evidence of the bloom on a satellite picture," he told me in 2009, but he said he couldn't remember whom or which satellite.) Wendy Williams, a journalist with the public radio show *Living on Earth*, called him "a smooth-talking . . . public relations man."

George had convinced several respected marine scientists that his business acumen would provide them with the money they needed to conduct expensive experiments on the high seas. But those relationships usually soured after the researchers grew uneasy with what they saw as a propensity for exaggeration, and what they felt were unseemly affiliations. (Planktos's main investor was Nelson Skalbania, a flamboyant developer from Vancouver who had been found guilty by the Canadian government in 1997 of stealing $100,000 in a real estate deal.) As he methodically alienated one leading ocean scientist after another, George became more dependent on his symbiotic relationships with the press. But he had a tenuous and tempestuous connection to both, complicated by his need for the legitimacy that each might afford him.

His uneasy relationship with authority became a particular problem when, soon after his public introduction at the National Press Club, the U.S. Environmental Protection Agency told him in a letter that if his boat flew under an American flag, he might require permission from the agency because it involved "dumping" into the ocean. George was defiant, making clear that if necessary, he would modify his plans to avoid the law. "The EPA has suggested to us that if we're not under the flag of a U.S. vessel, we're not subject to U.S. regulations. So if we were to register under a vessel that wasn't in the United States, we wouldn't be regulated by the EPA," George told a reporter in California, adding that more threats of "a long EPA process" would compel him to arrange the trip from overseas, perhaps out of an office in Budapest, where his company had some business in forests. "We have shipping agents in Central America working for us lining up vessels that might be able to assist," George told the paper.

Meanwhile, environmental groups began to renounce his company's plans. And reporters became increasingly skeptical that any top-notch oceanographers or biologists had lent their expertise to the company's efforts. George insisted that he had lined up "world-class scientists" to conduct experiments on a ship. But he would not name them when asked. "They have asked to remain anonymous because they're concerned that they're going to be victims of the same attacks that we have been," he told me. "I have way more people than I have berths." One scientist told me, in confidence, that George's claim was true.

Meanwhile, another company was exploring the idea of iron fertilization, but keeping a much lower profile. In 2005 George had met Dan Whaley, who had run a successful Internet travel business called GetThere before selling it for $750 million in 2000. Whaley was interested in starting his own "clean tech" company, as they say in Silicon Valley. He had a particular interest in iron fertilization, in large part because his mother, an oceanographer, was an expert in that area. (As a teenager, Whaley had served as an assistant on several scientific cruises.) Whaley met George in the Planktos offices for several days as the two considered doing business. Skalbania wanted Whaley to invest, and in exchange for Whaley be the CEO of Planktos. Whaley demurred. "He had a bad reputation among some very important scientists," said Whaley. Also worrying was George's position as head of a company called D2Fusion, which sought to sell what it called an appliance-size "cold fusion" device. "He was not the right person to do business with," said Whaley, who offered to buy the company outright instead of partnering with George. The two went their separate ways when George rejected the offer. Months later, four top oceanographers demanded that George remove their names from a filing to the Securities and Exchange Commission after he listed them as advisers without their permission. George blamed what he called a "mistake" on a subordinate.

"This guy is screwing up the whole field," Whaley recalls thinking. At that point, Whaley told me, he decided to make his own company. He called it "Climos." "Climos was a reaction to Planktos," he says. George soon after accused Whaley of stealing "intellectual property" from him.

The two firms had very different approaches to the controversial line of business they had chosen. In contrast to George's poisonous reputation among scientists, Whaley brought into his company a highly regarded scientist who knew as much about iron fertilization as anyone: his mother. Margaret Leinen was the consummate insider, the former director of the Geosciences Directorate at the National Science Foundation. That made her the P. Diddy of oceanography—she knew everybody, her phone calls were always answered, and half the ocean scientists in her Rolodex owed her a favor. She had assembled a scientific advisory board that included a former NSF director, prominent oceanographers, and Tom Lovejoy, a very-well-connected ecologist and environmentalist. She used connections among scientists and in Washington to make her company's message clear from the beginning: iron fertilization was unorthodox, sure, but it had to be considered.

And while Planktos lambasted the EPA, Climos was very public about its desire to work within the system. "If iron fertilization is to be done, it is to be done credibly and scientifically," said Whaley. Climos sought from the beginning to burnish its reputation among academic oceanographers. "Work with us," Leinen told a group of academic scientists at a meeting in Massachusetts in the fall of 2007 to discuss iron fertilization. George, by contrast, was invited to speak on a panel at the meeting but chose instead to stay in the audience, calling the meeting later a "kangaroo court seminar" featuring scientists "leading the conspiracy against Planktos." He later complained that he wasn't given a chance to "tell the truth" about his company. A few days after, Climos released to the public a "code of conduct" for their operations, calling for "transparent and open" experiments and "high standards" for any carbon credits they sold. Reached by phone, George indicated that his company supported the same ideals. George was "a little bit casual, off-the-cuff," said the University of Hawaii's Dave Karl. "Climos was the antithesis of that."

But Planktos had a head start—and cash. Riding a wave of positive press attention leading up to the Washington visit, Planktos's stock had risen to $2.56 by the end of the first quarter of 2007, giving the fledgling company a value of $91 million on paper. That allowed them to purchase their 94-ton ship, *Weatherbird II*. Morale was high as George's employees worked to refurbish the vessel with research equipment in a shipyard in Norfolk, Virginia. George took a photograph of a Planktos crew member "wearing all the firefighting gear, including the air mask, while holding a French maid feather duster," an employee named Melodie Grubbs wrote on her blog. "We are celebrating Chinese New Year on the boat, and have considered lighting our many expired flares. We finally have a stereo and are thinking about hosting a shipyard dance."

As the Planktos crew readied for their journey, the media couldn't get enough. The Discovery Channel broadcast a segment featuring George's young, mostly female crew taking water samples, examining them under microscopes, and wearing tight black T-shirts emblazoned with the company's globe insignia. Skalbania, George's investor, worried that green groups would protest *Weatherbird*'s mission. "We are the environmentalists," George told him.

On April 18, 2007, deckhands tending the yachts at a posh Fort Lauderdale, Florida, marina took in an unexpected sight: *Weatherbird* pulling into its slip, a handful of young female marine biology majors on board. "Those off-watch started early out on their tans out on the deck, I presume so as not to scare the sun-soaked Floridians with our paleness, when we pulled in dock," wrote Grubbs on her blog. "Our vessel and operations have attracted various peoples around the docks, I'm thinking reasons being one we're a cool crew and look fun to hang out with." George's plans for the company's journey, meanwhile, were falling behind schedule.

Delivery of scientific equipment was late, but George vowed to depart for the Galapagos via the Panama Canal in May.

For several weeks the crew waited for instructions from the boss. Meanwhile, from port, aft, and starboard came the desperadoes. Montréal-based nonprofit ETC Group complained in a May press release that the iron "nanoparticles" Planktos planned to dump would "foul Galapagos seas." Planktos "threatens our climate, our marine environment and the sovereignty of our fisherfolk," said Acción Ecologica of Ecuador. "Climate change should be tackled by reducing emissions, not by altering ocean ecosystems," a scientist with Greenpeace said. By June George's investors had sunk more than $3 million into the company, but the controversy had convinced some of them to withhold further funds. He was livid. "We are being swiftboated," George said.

After two months in the Fort Lauderdale marina, members of the crew became increasingly frustrated as their experiment faced an uncertain future. ("I was unable at the time to tell them that we were having trouble with our investors," explained George.) Grubbs blamed the "drama" caused by the emphasis on the pristine Galapagos in the press. While his crew contemplated jumping ship, George lashed out in public at his opponents. An editorial he wrote titled "I Am Not the Enemy" appeared in the *Ottawa Citizen* newspaper in June. "Why in a time when our beloved planet is in dire straits, would environmentalists turn on their own?" he asked.

But George was facing an attack that went at the very premise on which the whole project was based. In July, researchers overseeing a fifteen-hundred-scientist international ocean research consortium called the Surface Ocean-Lower Atmosphere Study published a one-paragraph statement questioning whether iron fertilization would "significantly increase carbon transfer into the deep ocean." In other words, could scientists on any experiment prove with certainty that the carbon their blooms had captured from the atmosphere was actually sequestered in the deep ocean after the algae died? Measurements taken from small-scale iron fertilization experiments have shown that growing algae can supercharge the growth of microbes that exhale nitrous oxide, a greenhouse gas more potent than carbon dioxide. In addition, algal blooms are hard to track, and the amount of falling carbon they shed is extremely difficult to measure. To untangle the mess either way one needs top-notch geochemists and physical oceanographers to start with, and devices called sediment traps that follow parcels of water up and down while collecting the carbon-rich "marine snow." Planktos had neither on *Weatherbird*.

George argued that his company would have to answer to authorities higher than his academic detractors. On July 17, in a ceremony held in Vatican City, a cardinal representing the pope accepted a gift of carbon credits from George, calling the Holy See "honored" to receive the donation. The following day, in Washington, George testified before a U.S. House of Representatives committee. Selling the carbon credits he'd claim at sea, he said, would require approval from UN agencies who would confirm the carbon measurements. "If we succeed, we will have created an industry. If we don't succeed, we will have created a lot of great science," George told the panel.

Meanwhile, the sailors were getting restless as *Weatherbird* remained in dock. In the late summer, as George was trying to raise money and advocate for his embattled company, his investors had nixed the Galapagos plan, and the company was mulling alternatives. So the *Weatherbird* crew drove rental cars to Fort Lauderdale to look for collaborators among Florida ocean scientists. Several were interested in joining the expedition at first, Grubbs told me, but after a day or two of gathering information about Planktos online, they demurred. One scientist, from Florida Atlantic University, was "very enthusiastic about working with us," she said. "But when we followed up a week later, he didn't want to talk about it." By September, after the company's stock price had fallen into the pennies, speculators unaffiliated with the firm launched a direct-mail advertising effort to hype the company's approach. "Global warming is a booming business," read the eleven-page mailer. "Planktos is solving a major threat to the health of our oceans. . . . This is YOUR opportunity to get rich from this undeniable trend."

On November 4, 2007, *Weatherbird*'s journey finally began. To lead the ship George hired two veterans of Greenpeace's ship, including Peter Wilcox, who George called "the greenest man in the ocean." But George kept the destination secret; instead of the Galapagos, the crew was heading to the Spanish-controlled Canary Islands, 150 miles off Morocco in the eastern Atlantic. There, some data suggested, the waters offered the right conditions for an iron fertilization experiment. George flew to the Canary Island port of Las Palmas, getting verbal approval for the ship to dock there, and arranged to purchase the iron and provide electrical hookups for the ship.

Wilcox piloted the ship to Bermuda to get supplies and fuel for the journey across the Atlantic to the Canaries. In Bermuda, authorities requested to enter the ship, searching for what they suspected was "toxic materials" on board, but found nothing. Then, in a stroke of *Gilligan's Island*–style bad luck, it turned out that *Sea Shepherd*, a ship piloted by radical pro-wildlife activists, happened to be in Bermuda at the same time. A spy from the organization pretending to be a tourist asked for permission to take some pictures on board *Weatherbird*. "We thought something was up," says Grubbs, "but we had nothing to hide." In a statement released afterward on its Web site, *Sea Shepherd* warned Planktos not to carry out its plans. "We are not Greenpeace. We won't be just showing up to hang banners and take snapshots," said the group, which has rammed and sunk nine whaling ships in its defense of the seas over almost three decades. (In 1979 the group had split off from Greenpeace, its rival organization, and sneered on a press release that George was "a former Greenpeacer.") George, for his part, said that members of the crew, and in one case their family members, were receiving threatening e-mails from opponents of the project.

Weatherbird managed to slip out undetected a few days later, and by early December, Wilcox directed the ship into Spanish waters about twenty miles from Las Palmas. He radioed for instructions, but the Spanish authorities warned him not to enter the port. He asked again, but the authorities repeatedly refused, accusing the company of planning to dump "toxic waste" into the ocean. *Weatherbird* then circled the area on low power as food supplies dwindled. The boat idled for days on end, circling the area outside the port. "By the seventh day I was getting worried for my safety," says Grubbs. "We had no fresh vegetables or fruit for days." The crew became adapt at catching mahimahi fish off the side of the ship, eating their catch raw—"sashimi style," as Grubbs put it, saving cooking fuel. A few days later, three crew members became sick, two violently ill with food poisoning, shaking and vomiting. George told the Spanish authorities that his team faced a medical emergency and needed to dock immediately, but they wouldn't relent. So with fuel supplies dwindling after a week of circling the Canaries, Wilcox decided to set course for the Portuguese island of Madeira.

From there, most of the crew flew home. The experiment was canceled. George blamed a "smear campaign by an environmental conspiracy" as the company's end came swiftly. In December of 2007 the company told the Securities and Exchange Commission that it had run up a deficit of $3.7 million. George and his partners managed to divvy up the assets without suing one another, selling *Weatherbird* to a company that monitored water around oil rigs. George started a new group, a nonprofit he called Planktos Science. In 2009, when I asked him to name his investors, he demurred. "We'll have iron in the water" that year, he said. Later that year, he told me that delays had "pushed the date back."

When Dan Whaley had become a climate entrepreneur, he was a Silicon Valley veteran with a lot of money and an appetite for risk. In 1995, at age twenty-nine, he had started a company called the Internet Travel Network with Bruce Yoxsimer, who he met in a tae kwon do class. The business's first office was in Yoxsimer's house in Palo Alto, California. Soon after, with only one thousand dollars in the bank, the start-up signed a $10,000-per-month lease for an office. "They got their first paycheck from their first client, American Express, within days of the rent due date and barely squeaked by," a local business journal recounted. Five years later Whaley sold the company, then called GetThere, and five years after that, he met George.

So it was a frustrating first few years for Climos. George had assumed the maverick role while Whaley and Leinen attended one international meeting after another to make the case for aboveboard geoengineering. That meant much more time spent giving presentations, lobbying, and pushing paper in various conference rooms than actually planning a research mission. The pair were a quizzical sight at scientific meetings: Whaley had a lazy nonchalance about him, leaning over during presentations with his arm out to the side. With a demeanor like a librarian and glasses to match, Leinen listened intently, taking extensive notes in a bound journal. Whaley called his mother by her first name, Margaret.

Two thousand eight should have been the company's best year. In March Climos secured $3.5 million from Braemar Energy Ventures, a venture capital firm and Elon Musk, a thirty-six-year-old Silicon Valley tycoon who ran companies selling private rockets and electric sports cars. They hoped next to raise the $10 million they would need for a research expedition to sell carbon credits "with a 2009 vintage," Whaley said. The company spent much of its revenue and many hundreds of hours preparing patent applications, environmental assessments, and scientific "methodologies" that might have allowed them to sell credits if science projects they conducted could have been shown to quantify the iron sunk by their experiments.

But even after Planktos had gone keel up, the ghost of the company continued to haunt their more polished geoengineering competitor. A particularly important event had occurred a month before *Weatherbird* headed for the Canary Islands in 2007, just as

Whaley and Leinen had attended a meeting of countries who had signed the London Convention, an international treaty on ocean dumping. There, diplomats from 82 nations would decide whether and how iron fertilization experiments would be regulated under the treaty.

Before the 2007 meeting, the issue had been one of many on the agenda for the convention, which covered everything from toxic sludge disposal to fishing waste. But Planktos and its devil-may-care attitude put the issue front and center among the delegates. At one point a triumphant—and to many, defiant—press release from Planktos was distributed at the meeting, much to the dismay of Leinen and Whaley. "Planktos sort of swayed the mood," an American diplomat told me. "People who might not have cared learned about the issue." And any distinction that American scientists drew between the two companies was lost on diplomats who might have known little about oceanography. "People mixed them up—their names [Planktos and Climos] were similar. They would say, 'There are American companies looking to do this.'" Planktos, the diplomat said, "really set a lot of people into a harder line against iron fertilization." At one point a lawyer working for Climos asked the American diplomat which countries supported iron fertilization. "It's just us and Saudi Arabia," the diplomat responded. At the meeting diplomats adopted a statement that said the science surrounding ocean iron fertilization currently is "insufficient to justify large-scale operations" and said they would set up rules to regulate it.

"We joked about whether Planktos was actually a nefarious plot to make sure that ocean fertilization never happened," said Climos employee Kevin Whilden. In May 2008, bureaucrats continued to target the controversial technique. Nations party to the 1992 Convention on Biological Diversity met in Bonn, Germany, and behind closed doors, with little input from scientists, passed a nonbinding statement that called on governments to forestall "ocean fertilization activities . . . until there is an adequate scientific basis on which to justify such activities."

That was irksome enough for many marine biologists, but simply mind-numbing was the exception the resolution made for "small-scale research studies within coastal waters." That made little sense to either scientists in favor of the experiments or environmentalists opposed. Experiments there would be fruitless, since naturally occurring iron concentrations were high, and therefore it would be very difficult to grow new algae blooms. And off the coasts, scientists might inadvertently grow poisonous algae that could harm beachgoers; this unlikely outcome was of little consequence in the open ocean hundreds of miles from any human settlements. (Even Canadian environmentalist Jim Thomas, a leading opponent of iron fertilization, told me later that "the language about coastal waters was a mistake," though he supports the ban that the treaty represented.) In the fall, the London Convention issued a statement that ocean fertilization activities should include "legitimate scientific research only," which, though not explicitly, seemed to rule out commercial activities and cut the company's business plan out from under them.

It was at about the time of the London Convention statement that Leinen and her son decided that their company would have to abandon its position that it would try to sell carbon credits based on their experiments. "It was just too controversial an issue," Leinen told me. Instead, Leinen founded a nonprofit organization, the Climate Response Fund (CRF), to raise money for geoengineering, hiring a well-connected fund-raiser in California named Danielle Guttman-Klein to do so. Headquartered in Arlington, Virginia, the organization had an unclear connection to Climos. Leinen said the two were completely separate, and that she had no financial stake in the company. Whaley told me in the spring of 2009 that "appropriate conflict of interest controls were in place" but that the organization would financially support any iron fertilization experiments that Climos helped organize. "CRF will fund the researchers directly," he told me. "Climos will handle the logistics." Geochemist Ken Caldeira questioned whether Leinen's group was appropriate to host discussions on geoengineering, as it planned a

March 2010 conference on planethacking research and regulation. "There's a perception that you've got a fox in the henhouse—for-profit companies or their nonprofit surrogates looking at governance of geoengineering." But the event drew prominent scientists from around the world.

I asked Russ Lamotte, a lawyer who worked for Climos, if it was frustrating that the company had played by the rules and still found its prospects limited. "You have to wonder what would've happened if we had gone about this in a very different way instead of having to be reactive to the wave of negativity," he said. But Whaley was optimistic about the future of the iron fertilization business. He said he was encouraged by the new international rules being developed by the London Convention to regulate fertilization experiments at sea, which he said would allow the iron fertilization community to "put the specter of Russ George behind us" for good. "Philanthropy, government grants, we'll have a way to make it work. They're always going to be some ups and downs," he said. "We're actually a couple of really good people trying to push the eight ball down the field."

IN THE 1950S, SOVIET ENGINEERS EMBARKED ON ONE OF THE largest water projects of all time. The goal was to provide water to the desert plains of the Soviet Union. The nation needed cotton, planners decided, and to get water to the dry areas around the Aral Sea to grow it, engineers carved hundreds of kilometers of canals to divert water from the two rivers that feed it. The project diverted billions of gallons of water from the Syr Darya and Amu Darya rivers, and the growing superpower's cotton exports skyrocketed.

At the time the world's fourth-largest lake, the Aral Sea began to deteriorate almost immediately. In the 1960s it became clear that the sea was shrinking as a combination of evaporation and runoff from agricultural areas began to destroy the lake. By the 1990s, the lake lost 80 percent of its volume, and nearly 9 million acres of lakebed were exposed. All twenty-four of the sea's native fish species went extinct. Rusting fishing boats sat on the sand of what was once a thriving cannery at Muynak. The exposed lakebed allowed dried fertilizers and pesticides to get whipped up by the winds, and in 1993 the United Nations estimated that the death rate from respiratory disease in an adjacent province in the northwestern corner of Uzbekistan was among the highest in the world. Cotton yields dropped as the salt spread.

In 1999 the World Bank began funding a rescue mission to revive and regrow a northern section of the lake by building a 13-kilometer dike. Seven years later, its waters had risen by several meters, and *Science* called the project "one of the biggest reversals of an environmental catastrophe in history." But, the magazine noted, new interest in the waters, by farmers in Afghanistan, threatened the fragile lake once again.

Victor's Garden

For twenty years, oceanographer Victor Smetacek has been captivated by the enigmatic story of the Southern Ocean, the 8-million square-mile expanse that rings Antarctica. It's a desolate vastness at the bottom of the world, whose latitudes sailors long ago dubbed the Roaring Forties and the Furious Fifties. But despite its cold temperatures, the ocean would seemingly offer ideal conditions for marine life. Full of nutrients, its waters are fenced in by a constant eastward flow that pushes a hundred times as much water as Earth's rivers. Smetacek calls the result a set of "merry-go-round ecosystems." They continually recycle nutrients such as nitrogen within their ecological niches, anchoring a food web of sea creatures that include everything from microbes to the largest beasts on Earth.

Sea life is for the most part scarce these days in the Southern Ocean, however. Smetacek considers most of it "a relative desert." The iron hypothesis suggests that what's missing is iron, previously delivered by wind as a component of dust from land. As a result, during the last ice age, 15,000 years ago, some scientists believe that enormous blooms of plankton such as algae blossomed all over this stormy sea. As they died or were eaten, the story goes, they took billions of tons of carbon with them to the seafloor. That reduced the amount of carbon dioxide in the atmosphere, lowering the greenhouse effect and cooling the planet. Scientists don't know how that happened. Or, for that matter, whether the hypothesis is more than a hypothesis, or less, and just how the once-vibrant ecosystem of an ocean Smetacek calls a desert had affected the world's climate.

Even before Smetacek understands this cold, wet desert, he wants to explore altering it radically. So he spent three years, starting in 2006, planning an unprecedented experiment in that desert to try to make it bloom. Like Russ George and Dan Whaley, Smetacek believes that restoring the iron could make the Southern Ocean a crucial repository for 1 billion tons of CO_2 each year. If George was a defiant geoengineer and Leinen a cautious one, Smetacek was insouciant, with lively eyes, almost reveling in the wildness of the idea. "This is like turning all of Siberia into mammoth land," said the oceanographer, who works at the Alfred Wegener Institute for Polar and Marine Research in Bremerhaven, Germany. "The Southern Ocean is now an interglacial ocean. We are talking about making it a glacial ocean."

When John Martin suggested the iron hypothesis in 1990, one political cartoonist lampooned the idea, depicting a scientist in a cartoon boasting that the concept would allow scientists to "turn nature on and off like a beer spigot." But Smetacek thinks of the approach as ecologically elegant, if whimsical. And yet as he and forty-eight other scientists set out by ship from Cape Town, South Africa, on a warm evening in January 2009, the fanciful idea of re-creating the ancient Southern Ocean was becoming real. As passengers aboard RV *Polarstern*, a German research ship, they were setting off to conduct the largest geoengineering experiment attempted to date.

The Indian and German governments had cosponsored the $4.5-million, sixty-nine-day experiment, which Smetacek had named LohaFex. "Loha" means *iron* in Hindi, "Fe" stands for *fertilization* (and happens to be the chemical symbol for iron), and "x" is a suffix oceanographers traditionally use to christen their experiments at sea. Adjacent to the ship's helicopter landing pad sat two metal shipping containers in which were stored thirteen hundred bags of iron sulfate, forty tons in all, the same stuff gardeners used to treat soil. They planned to use the iron to fertilize the sea. Their target, what Smetacek was to call "our garden," was a three-hundred-square-mile patch of algae they'd grow between South Africa and the Strait of Magellan. To track the experiment in such turbulent waters, the scientists would carry it out in the center of a naturally occurring whirlpool, dozens of miles across, known as an eddy. There, relatively stationary waters would allow the scientists to track the birth, life, and death of the algae bloom they hoped to grow.

Aboard perhaps the finest research vessel in the world, Smetacek and his team were certainly well equipped for the task. *Polarstern*'s 387-foot-long double hull could easily deflect waves or small pieces of ice, known as growlers; a dozen laboratories were connected by a computer network; two large cranes loomed above the main deck; the crew kept the sauna working; four engines provided sufficient force to break pack ice if need be; and when deployed, stabilizing fins maintained the ship's balance. As on other ships, the biologists attached tennis balls below certain equipment to stabilize it. But compared to other vessels they found *Polarstern* particularly stable, allowing them to manipulate samples under the microscope with particular ease even if the captain, twenty-year veteran Stephen Schwartz, was caught broadsides to the waves—which he rarely was. ("They have beer on board," an American oceanographer told me jealously. Alcohol is not allowed on U.S. research vessels.) Smetacek had assembled a stellar team: physical oceanographers who would plot and predict the movement of water masses, geochemists specializing in all manner of chemical components of the ocean, experts on sea creatures ranging from bacteria to fish. The ship's crew included a weatherman and a medical doctor who also was a dentist.

Smetacek and Wajih Naqvi, an Indian biochemist with a more reserved, cautious demeanor, led the scientists on the mission. But there was little question that this was Smetacek's endeavor—the pinnacle, in some ways, of a four-decade career in which he had spent an accumulated year and a half total on the stately vessel. Much of that time he'd led smaller fertilization experiments. "We have to get away from the old thinking, which is that we can simply observe these ecosystems to find out what's going on," he said. The spirit of the Blue Team. Oceanography was full of "ho-hum" observational papers, he told me; to truly understand this mysterious but vital part of the world, he wanted to prod it, bring it to life. "We don't know what's happening, and unless we go in and get our hands dirty we're not going to find out."

As a boy Smetacek explored the foothills outside his town in the western Himalayas, near Nepal, hunting, fishing, bird-watching, and collecting butterflies. At night, by the light of a kerosene lamp, he read stories of the legendary Jim Corbett, who had killed tigers and leopards that ate men in the villages of the area. But the ocean soon captured his imagination. "My father had been a sea-farer before arriving in Calcutta in 1939 and his romantic attachment to sailing ships strengthened my resolve to turn to the sea," he would write as an adult. Rummaging through old copies of *Reader's Digest* as a teenager, he found an article titled "Bread from the Sea," about the potential to provide food to the hungry from algae. "Can you expect one day to buy algae in your local food store? Quite possibly," he read.

The career Smetacek built combined oceanography with a variety of other disciplines, earning him a reputation as somewhat of a rebel. ("Territorial males hate to be called homodisciplinary," he wrote.) In 1992 he published an article in *Nature* about the surprisingly common phenomenon that some left-handed people, including Lewis Carroll and possibly Leonardo da Vinci, can write in mirror script. His ideas about gravity spanned papers on the human subconscious sense of balance, primate evolution, and the role of sinking particles in the ecology of algae blooms. He saw those blooms as an impossibly dynamic microscopic ecosystem where microorganisms zip around at speeds many times faster, relative to their size, than greyhounds. "Imagine yourself in a light forest looking upwards, seeing in your mind's eye only the chlorophyll-bearing cells of the canopy floating in mid-air, free from the attachment of leaves, twigs, branches and trunks. Now forget the forest and the trees, and see only blurred clouds of tiny green cells obscuring the blue sky beyond. You are looking at a phytoplankton bloom," Smetacek wrote in 2001.

During LohaFex, Naqvi provided geochemical rigor to augment Smetacek's almost uncanny ability to discern from instruments what was happening below the waves. The two scientists had dreamed up the project over fish and beer—Smetacek drank; Naqvi, a Muslim, didn't—at a sea-themed Bremerhaven restaurant five years before. "Did you know that the place in the body with the second most neurons than the brain is the stomach? That's why we call it 'our gut,'" Smetacek told me. "I think it comes from our sense of balance. I have a feeling about iron fertilization in my gut."

Geoengineering a poorly understood ecosystem such as the Southern Ocean was an uncertain but tempting proposition. On one hand, computer modeling suggested in widely cited papers that the Southern Ocean could take in 1 billion tons of CO_2 each year if scientists were audacious enough to fertilize virtually the whole thing. And yet there was so much the scientists didn't know about this mysterious, ignored place. A small sampling: What causes so much carbon to get sucked out of the air and brought down into the deep part of the ocean? What kinds of algae, the main plant of the ocean, grew where, and why? What precise role do bacteria, or the tiny sea creatures called zooplankton, or fish play? Talking about hacking the Southern Ocean is like a scientist proposing to build genetically modified trees in 1925, before scientists understood how moss, rain, and insects interact. Smetacek knows it's a daring idea. But the climate crisis might require such daring. That's just the quandary of this era, the Anthropocene.

• • •

The smaller iron fertilization experiments Smetacek had previously led were highly collaborative because they required constant decision making as to where the ship should go, when and where to add iron, and which measurements should be taken and when. Smetacek lived passionately for the experiences, providing infectious enthusiasm and making the scientists aboard *Polarstern* feel for the most part listened to and part of the team. It was well known among his partners and subordinates, however, that his scientific imagination sometimes could make him a little scatterbrained. And his optimistic outlook occasionally left him vulnerable to naiveté. "He's a great scientist, but a little disorganized," one American oceanographer said before LohaFex. "He can be crazy," was how Maria Hood, a science official with UNESCO, put it.

Still, he knew the storm that had engulfed Planktos. Nonetheless, he expressed his confidence in the morality and potential of the experiment in his genuine though slightly exaggerated British formality. "As your co-Chief Scientist I assure you that you will have an enjoyable time and take home a memorable experience working on board this sturdy vessel," he wrote in an e-mail to the participants. He described with typical flair the camaraderie he hoped to experience: "Since it will be Carnival time, perhaps not only the Goanese will appear in fancy dress?" he wrote. "We shall have to watch our weights. The gym is open 24 hours but the swimming pool will be filled only when the weather is calm." Toward the end of the letter he admitted, almost as an afterthought, "Iron fertilization is controversial," though he noted that he and Naqvi had secured the "blessings of our respective Governments" before casting off.

Those blessings had not come easily. For more than a year Smetacek had made what he thought would be sufficient preparations to assure that an experiment he knew would be controversial would still happen. At times it wasn't clear that it would. The timing was brutal. He had planned the largest ocean fertilization project to date just as the international regulators, environmental groups, and international law experts were moving to regulate the technique, riled up in no small part by Russ George's bravado. In May 2008, only seven months before LohaFex, the Convention on Biological Diversity met in Bonn, Germany, and agreed to restrict iron fertilization experiments to "small-scale scientific research studies within coastal waters." In what amounted to horrible luck for LohaFex, Germany's minister of the environment happened to be leading the negotiations. Along with most oceanographers, Smetacek called the ruling nonsensical. But the attorneys he asked, including those within the German research ministry, told him that the wording was "not legally binding." It was not a change to the treaty, they said, but rather actions that the treaty body had agreed on, and had not been approved by individual nations party to the agreement.

The scientists were armed with reams of other paperwork. The British government required an official review because Smetacek wanted permission to collect biological samples near waters controlled by the United Kingdom near South Georgia Island, east of the Strait of Magellan. That got the German Foreign Ministry involved, which shared the issue with the Environment Ministry. Two months before the journey, Smetacek said, a midlevel Environment Ministry bureaucrat told him that LohaFex could proceed, but that the scientists should do a systematic risk assessment beforehand. "What the hell do you want in a risk analysis?" Smetacek told me he thought. "I don't know what I would put in a risk analysis. There is no risk." Lawyers at the Wegener Institute agreed that he could ignore the request. "It was a yellowish-green light," Smetacek told me. "The Environment Ministry had no authority over us anyway."

But days after Cape Town disappeared behind *Polarstern*'s prodigious wake, political clouds gathered. Someone posted on a bulletin board in a lounge a South African news story that called the research vessel a "rogue ship" that had "slipped out of Cape Town harbor to conduct a controversial climate change experiment." The article's contention that the expedition had violated international law caused considerable dismay among the scientists. The younger Indian ones were certainly burdened with sufficient anxiety to begin with: many had never been on a research expedition before, or even close to this far south on the globe (let alone on a mission to grow a bloom of algae as large as sixty-seven thousand football fields and monitor the ecological response for months in close quarters under harsh conditions). Now they had been branded naval miscreants. The turn of events was particularly disheartening for Smetacek, who was hardly considered a reckless profiteer or environmental renegade by his colleagues. He didn't see geoengineering as a quick fix. "The most important thing is to curb emissions," he'd said repeatedly before LohaFex. If the technique proved to sequester carbon dioxide, he felt the United Nations should conduct iron fertilization projects, not private companies.

And yet environmental activists inside and outside Germany lobbied contacts they had within the German government to halt the experiment. "We need strong, enforceable rules to prevent rogue geoengineers from unilaterally tinkering with the planet," said Canadian environmentalist group ETC. The German Environment Ministry wrote the Science Ministry, which sponsored the expedition, asserting that the research was "undermining Germany's credibility and pioneering role" in leading the biodiversity convention and that the mission had to be canceled. The last line of the letter icily informed the Research Ministry, as though it didn't know, that *Polarstern* was "already on cruise to the South Atlantic." On the fourth day of the expedition, as the political heat rose, Smetacek's boss in Germany ordered him to put the experiment on hold. He was to prepare, at sea, the risk assessment document they had previously avoided writing.

Smetacek tried to keep a brave face, but he mostly retreated to his cabin "so as not to concern the others," he told me. But the scientists knew exactly what was happening. "Everything seemed to be falling apart around us," said Navqi. The delay could shave precious weeks of observation off the experiment, they feared, or worse, and the experiment might even be canceled. The crisis sentenced Smetacek to days on end on his computer, battling on the political front. Using the ship's satellite link, he sent dozens of e-mails to scientists, government contacts he had, and friends, urging each of them to do what he or she could to save his experiment.

A forty-two-point questionnaire submitted by the German Green Party required his attention. He took days to write a twenty-three-page, detailed risk assessment, working long into the night as the ship steamed past icebergs and diving albatrosses, heading truly to the middle of nowhere.

The assessment emphasized with little subtlety how Smetacek believed the experiment would comport with the Biodiversity Convention's call for a small, "coastal" experiment. Years before LohaFex, Smetacek had repeatedly said how "larger-scale, longer-term" experiments like the one he was planning were just what scientists needed to better understand if and how iron fertilization could help suck in carbon. Yet in the document Smetacek repeatedly mentioned the "small magnitude" of the experiment. By adding iron, the concentration of iron in the water would be roughly one twenty-fifth the concentration of natural iron found in coastal waters, he noted. Despite the fact that the experiment would be happening hundreds of miles from any shore, he also repeatedly mentioned the "coastal" nature of the water: various chemical isotopes would be "measured as a proxy for coastal influence," their scientists would quantify "the presence of typically coastal phytoplankton species," and the water they would fertilize had, yes, once been adjacent to a coast. (World Wildlife Fund's Stephan Lutter, a former student of Smetacek's, called the effort to portray the experiment as a "coastal" effort "really outrageous," especially in light of the Biodiversity Convention's intent.)

It was bordering on irresponsible, Smetacek said in interviews, not to further examine the technique's potential for mitigating climate change. Each year human society spews out more than 8 gigatons of carbon emissions, with the ocean thought to take up a quarter or more of that total. (Phytoplankton bind as much carbon dioxide in the ocean as land plants do.) Why not think of ways to enhance that potential sink? Smetacek and Navqi noted that in total the oceans store an estimated 38,000 gigatons of dissolved carbon in various forms. "So adding a few hundred gigatons [over time], if adequately diluted, is not going to make much of a difference," they wrote. "Not considering CO_2 removal now is tantamount to not bailing out water pouring into a sinking ship," Smetacek said. He also hoped that the experiment could test whether algae growth could indirectly boost stocks of a shrimplike creature known as krill, a crucial part of the Antarctic food web that feeds decimated populations of great whales.

News stories and bloggers tracking the controversy kept LohaFex's scientists jittery as they set up their laboratories and the ship steamed west, in the direction of Tierra del Fuego. Meanwhile, another crisis cropped up: they were having trouble finding an appropriate eddy in which to conduct the experiment. Within the enormous and unpredictable boundaries of such whirlpools, there was a semblance of a closed environment, it turned out, for weeks on end. In previous years, the scientists had found an appropriate candidate at 50° south latitude, roughly 2,900 miles from Cape Town. Within the confines of the eddy, scientists could try to keep the bloom from breaking up. A French physicist, reporting daily by e-mail from Paris, was tasked with determining via computer modeling the stability of the eddies in the area. The Oracle, as he became known, reported that the eddies in the area where Smetacek targeted—a site where the shape of the ocean floor apparently encouraged the formation of eddies—were prone to falling apart, making them bad candidates for the experiment. *Polarstern* encountered a more well-defined eddy along the way, but samples revealed that its waters contained low levels of silicic acid.

That was a problem because Smetacek was trying to grow algae known as diatoms, which built their protective glass shells of silica, obtained from the silicic acid dissolved in seawater. A handful of other experiments, each on a much smaller scale, had grown phytoplankton patches in the Southern Ocean; silicon-requiring diatoms had been the main type of phytoplankton that grew, and in one case appeared to sink precipitously. It wasn't clear what would grow if the scientists fertilized an area with low silicic acid available in the water.

The ship steamed about a thousand miles west to investigate a few eddies. Two were weak, confirming the Oracle's premonitions.

Then the ship explored some icebergs. Finally a promising eddy appeared on the daily satellite readings, and investigation showed plenty of silicic acid in the water. But as they discussed the possibility, the ship's captain shouted at Smetacek that the center of the whirlpool sat within British-controlled waters, and conducting the experiment there would have required permission from the British government. Given the controversy around the cruise at that point, the chances of obtaining the paperwork seemed remote at best—and the captain told Smetacek firmly that he didn't want the additional hassle. Another eddy looked intriguing some 575 miles south—but the German government had forbidden them to conduct any experiments south of 50°, due to an environmental treaty. "We were very frustrated in many respects, but we knew we couldn't push it any further," said ecologist Philipp Assmy.

So the scientists decided during one of the daily science meetings to head back to the first eddy, silicon or not. Smetacek later regretted not having closely looked at the map of silicon distribution. "I have to admit I forgot the silicon," he told me.

On January 26, three weeks into the cruise, *Polarstern* settled into the center of the first eddy, awaiting news from Berlin. The day began with a flag-hoisting ceremony to commemorate India's Republic Day, for which the Indian scientists sang their national anthem. A cultural event followed, featuring Indian vegetarian dishes, Indian flag buntings, songs, poems, and dances rendered in ten different Indian languages, from Hindi to Tamil. Amid the festivities Smetacek got the call: the reviewers of the environmental assessment had given a thumbs-up to the experiment, and the Research Ministry gave Smetacek permission to commence. During a break in the middle of the day, wearing a red traditional Indian shirt called a *kameez*, he told scientists assembled on the main deck that the experiment was on.

It took volunteer teams of scientists working for hours, spread out over two days, to haul more than four hundred bags of iron sulfate powder, each 55 pounds, to mixing basins on the main deck. The powder could irritate the skin, so the scientists wore protective suits and masks as they mixed the material in large basins with water. A tube ran from the tanks to the back of the ship, where it released a trail of iron solution into the ocean. For more than a day, the ship began a series of tight turns, laying down the fertilizer in increasingly large concentric circles around a buoy that marked the center of the patch.

"Our experiment will show us how the anemic plankton reacts to this 'manna from heaven,'" Smetacek had written in one of his colorful ship logs that he liked to e-mail back home. Previous experiments showed that dropping that manna on a patch of water with sufficient silicon—as well as other key nutrients—grew massive diatom blooms. Leaving Cape Town, that had been the goal of the experiment, since big blooms could sink and bring with them carbon. Without silicon, the LohaFex scientists were conducting a slightly different experiment, since they knew the diatoms wouldn't grow. They were mostly growing a tiny roundish algae known as *Phaeocystis*, which have no shells and several flagella. Dead remnants of *Phaeocystis* blooms often form thick, nasty yellow scum along the coasts of rivers and oceans around the world. ("*Phaeocystis* has a bad reputation in the minds of many people. However, this view is subjective and unjustified," Smetacek wrote in another of his e-mails.) The scientists were disappointed not to grow diatoms, but attempting to cultivate a *Phaeocystis* bloom this large had its own appeal. Marine biologists have found big blooms of the tiny species along the coast of Antarctica, but they didn't know how important the species was in the global scheme. Algae come in all colors, but the scientists would be unable to see the growing bloom with the naked eye from their ship. Their "sensing organs," as Smetacek called them, included sampling bottles that opened at various depths, lowered from the side of the ship. Using nets, the scientists caught fleck-size life, which they observed under microscopes. A camera recorded ephemeral black-and-white snapshots of the microscopic chaos below the patch.

Roughly a day after beginning the fertilization, the scientists measured a gratifying effect: the algae in the water were conducting photosynthesis twice as fast as before. The ship began surveying the patch in a crisscross pattern while towing a device that recorded vital signs such as temperature and salinity.

When the meteorologist on board warned of bad weather coming, the scientists decided to steam north out of the eddy while the storm passed. When they returned it was a bit difficult to find the center of their bloom, in part because GPS-equipped marker buoys they had dropped into its center had been pushed off course by ocean currents or winds and were executing a series of maddening loops, plotted on the computer screen. Meanwhile, various currents were ominously stretching the eddy's edges. By the second week into the experiment, the scientists were spending much of their time just keeping track of the location of the experimental patch they had grown. "Each time we had to run away and come back with suspense," marine biologist Marina Montresor said. "It's like a game, a detective game." With its humming engine and hulking cranes, *Polarstern* seemed huge to the scientists when they went out into the night's cold air. But from above it was only a white dot on a patch of invisible algae that itself was a dot within a slowly rotating whirlpool 62 miles wide, in rollicking seas, at the bottom of the world.

Each evening after dinner the scientists discussed the ongoing experiment in meetings held in a wood-paneled conference room maintained with German fastidiousness—no food or drinks allowed. There they shared data "hot off the instruments," as they liked to say. And the perpetrators and victims in the grand drama unfolding beneath the patch came slowly into focus. First the fruits of their garden were sprouting their shoots: a relative handful of diatoms had grown in response to the iron fertilizer, connecting together to form snowflakelike shapes, four ten-thousandths of an inch across or smaller. Crowding them, almost trembling, were the minute round *Phaeocystis* cells. They were hiding, as it were, from the onslaught.

Swarming about amid the growing planktonic feast, the predators attacked. The main ones were two species of sea creature known as copepods—one about the size of a mosquito, and the other, a flea. Most had migrated up the water column from below, where they tended to hibernate in the nearly freezing water between feeding seasons.

In response to the iron, the total amount of material in the algae patch, known as biomass, had roughly doubled. That's the kind of growth that previous experiments growing the hardy diatoms achieved in their blooms. When the omnipresent clouds cleared for a few hours, a satellite managed to capture what amounted to a chlorophyll snapshot of the area, revealing a gratifying signal in the precise shape of the eddy they had fertilized. The scientists decided to provide a second round of fertilizer, dispensing 10 tons of iron solution back into the patch to see if the *Phaeocystis* would respond.

A week or so later, scientists began to realize that their patch was failing. When they're healthy, *Phaeocystis* forms colonies of tens of thousands of cells to protect itself from predators, growing big blooms. But four days after the second fertilizer dose, the largest colonies the scientists could find in the patch were balls of eight cells, and even those were rare. The zooplankton were having a field day. "The zooplankton swept the water column clean," marveled Assmy, the ecologist. But not entirely, he noted. Here and there *Phaeocystis* algae remained. Without a big bloom, however, the possibility of sucking much carbon dioxide out of the atmosphere was lost.

A combination of politics and bad luck may have thwarted Smetacek's original goal, but two mysteries that emerged about a month into the experiment kept the scientists' attention regardless. One was the question of where the iron had gone. The other was why the algae hadn't been completely obliterated.

A Spanish chemist named Luis Laglera emerged as a crucial scientist as the researchers tried to understand what happened. Laglera was perhaps the researcher with the hardest task on the ship—to measure, painstakingly, the iron concentration in the various segments of the ecosystem. On a ship itself built of iron, his work demanded incredible fastidiousness. The big bottles the crew lowered into the water to take samples were made entirely of plastic, the frame that held them was coated with rubber, and the whole apparatus was connected to a crane on board with a special Kevlar cable. He performed his analyses under a microscope for hours on end in a ventilated room underneath a small plastic tent into which visitors could not enter; he played Spanish punk rock to pass the time. "I was the weird guy in the bubble," Laglera said.

Instead of battling contamination, however, Laglera encountered the opposite problem. He could barely find evidence of the metal fertilizer anywhere. The iron they found in algae samples they took was only a portion of the total amount of fertilizer they had released. Where was the rest of the iron? Laglera couldn't find iron in the water below the patch, or as particles suspended near the surface or gobbled up by organisms. What little his colleagues actually saw of him—he generally grabbed his dinner and headed back to his lab—suggested he was getting frustrated.

The mystery was connected to the behavior of the algae. What, or who, was continually feeding the *Phaeocystis* colonies with iron fertilizer? When nature provides the iron from coasts or icebergs, as part of dust, the particles tend to coagulate with one another and sink—organisms either get their meal or they lose it. So two weeks after the second iron dose, the small but persisting specimens of algae couldn't be living off the original slug. Somehow the system was recycling the iron that was keeping it alive—a phenomenon scientists hadn't seen before.

A mix of desperation and excitement characterized the evening sessions. Might the copepods be recycling the iron? They certainly seemed to eat a lot more than their bodies needed. The assumption had always been that the excess amounts of dissolved iron leached out of the bodies of the frenetic creatures as they gorged, where it would quickly coagulate and fall away from the surface. If that happened, it would be unavailable to the algae.

Was there iron in particle form stuck in the fecal pellets? In the final weeks of the experiment Laglera began a marathon effort to try to find out whether the copepods were recycling iron particles in their waste. For fifteen hours a day he holed up in his tent, manipulating the minute pellets with ultraclean tweezers under an ultraclean microscope. "He was completely pale," said Assmy.

During someone else's evening presentation, as part of a comment from the audience, Laglera announced that he had found iron in the fecal pellets. "That was a major finding, and very exciting," Smetacek told me later. "It's like agriculture, what these copepods are doing," Smetacek told his colleagues. The copepods were tending to their own garden. Forty-eight scientists and forty-seven crew members had traveled sixty-nine days and 12,488 miles to try to reshape a little nature in a way that might help humanity. But tiny sea bugs had their own plans to hack the ocean, and they were better at it.

Once home in Germany, Smetacek was asked to prepare a press release describing the results of the controversial experiment. "LohaFex has yielded new insights on how ocean ecosystems function. But it has dampened hopes on the potential of the Southern Ocean to sequester significant amounts of carbon dioxide," read the statement. "There's been hope that one could remove some of the excess carbon dioxide," Smetacek told the BBC. "But our results show this is going to be a small amount, almost negligible." *Discover* magazine's blog, echoing other press reports, called LohaFex a "bust."

But two months later, Smetacek recanted. He was speaking by telephone from his house in Bremerhaven, Germany, alongside a forest that sat next to the autobahn. It was afternoon my time, late at night for Smetacek. I asked about the press release and the statements he made after the cruise. "We were under political pressure to release that kind of statement," he said. "We have to continue to look at iron fertilization; it's just too important not to."

For all of Smetacek's abilities as a scientist, it occurred to me that geoengineering required a different set of skills. A thicket of international rules and German environmental politics had almost ended his experiment at sea before it started. He had failed to ensure that LohaFex would be able to find a spot with sufficient silicon so they could perform the experiment they had planned. And now, back on land, the powers that be had forced him to back a statement that would certainly dim hopes for more iron fertilization experiments. His savvy was of the personal variety, not the political; he was too brash and honest and not, perhaps, Machiavellian enough to be an effective geoengineer.

He wasn't much of a time traveler, either, or at least not a very precise one. Smetacek had wanted to use LohaFex to convert a small patch of the sea into the Southern Ocean of 15,000 years ago, the last glacial period, when there was plentiful iron blown in from dust to feed diatoms. He had provided the iron, but without silicon, the algae lacked material to build their shells to protect themselves from predators. Instead, it was as if he'd gone back hundreds of millions of years before, when the diatoms had yet to evolve to play such a big role in the ocean's ability to sequester carbon. Back then, it was unarmored algae like *Phaeocystis* that grew in the sea, vulnerable to predators and bad at sending carbon from the sky to the deep. By growing sea plants with iron but without silicon, it was the world of the vulnerable algae that Smetacek and his team had explored.

But Smetacek was more interested in talking about what he had found in the Southern Ocean during LohaFex and not what they had failed to do. After the rule of the unarmored algae at sea began to wane, Earth began a long, gradual cooling period that shaped the modern climate. "Rapidly evolving diatoms began to make big blooms in the oceans," he theorized. Armed with evolved silicon shells, diatoms foiled their predators and grew big blooms. After that they would die, pulling the carbon they had sucked in from the atmosphere, as well as nutrients, with them to the seafloor. The planet's temperature would eventually fall accordingly.

LohaFex showed that even with sufficient iron, big blooms of unarmored algae wouldn't readily grow, since their predators would devour them before they had a chance to form colonies. But the diatoms could survive the onslaught. So, said Smetacek, the result strengthened the counterintuitive idea that minuscule diatoms had a big role in cooling the planet since the days of the dinosaurs, 100 million years ago.

But Smetacek, a night owl, was just getting going. Even though the experiment didn't deal much with diatoms, it did shed light on their secrets. As the diatoms had become dominant in large swaths of the ocean, and they fell if they weren't eaten, wouldn't the nutrients in the water, like iron, get depleted and doom later generations that would need them? LohaFex had suggested for the first time that the algae's predators, copepods, had co-evolved to help them by continually fertilizing the surface with recycled iron in their waste. In turn, the copepods needed to keep nutrients on the surface for the copepods' young, who would need algae to eat after their parents died. "It's like antelopes and cows who have a similar relationship with grass," Smetacek said.

He was going on about the copepods in a rather scatological way. I recalled his introductory letter to the LohaFex participants: "You should not dispose of any object not meant for the toilets, that operate under suction, in them. Experience has shown that even a rubber band or string can lead to blockage of the suction pipes and nasty work for the engineers." "This organism is obsessed with its feces," I remarked, meaning the copepods, not him. "We have not been obsessed with our waste problems and that's why we're in this mess," he responded, much more soberly than usual.

THE LARGEST TROPICAL LAKE IN THE WORLD, LAKE VICTORIA PROVIDES food and water to hundreds of thousands of people in Uganda, Kenya, and Tanzania. In the 1950s its waters held more than three hundred species of fish found in no other lakes, and ecologists from around the world flocked to its shores to study the ecosystem. But as demand for fish among people in the area rose, the populations of small native species, including tilapia and labeo, grew scarce. Fishermen wanted a bigger, more commercially viable catch.

Foreign scientists working in Uganda thought that a larger fish, called Nile perch, would feed on the smaller native fish and provide a far greater haul for the fisheries. Since there was concern that they might wipe out the native species, experts conducted trials in nearby Kyoga Lake and Nabugabo Lake. But before the results could be analyzed, a few Nile perch were found in Lake Victoria in 1960. Two years later, officials introduced thousands of them to Lake Victoria. Local fisheries subsequently enjoyed a fourfold increase in their yields from the extra fish.

But the ecological price, compounded by overfishing and pollution, was staggering. Two thirds of the native species were driven to extinction; Lake Victoria's depths became ravaged by vast, low-oxygen dead zones; and algae blooms made the waters murky. The ecological calamity has meant economic disaster for local populations that generally could not afford the perch, which was more expensive than previous varieties. Les Kaufman of Boston University called the turn of events "the greatest vertebrate mass extinction in recorded history."

The Sky and Its Reengineer

Scientists dream up ideas, but engineers make them work. And so if geoengineering is to be done, it will be done, aptly, by engineers.

Perhaps the biggest opportunity in engineer Stephen Salter's life came in 2003 when he met a scientist named John Latham, who needed help making cloud-brightening ships that wandered Earth's seas. Six years later, the Scottish engineer and his colleague would get a chance to receive a scientific grant from Bill Gates to build ships to stave off global warming.

Salter was imaginative and grave, a Member of the British Empire for his inventions harnessing winds and oceans. His patents formed the bases for a number of start-up companies. But his successes had neither dulled his imagination nor dampened his willingness to suggest outlandish ideas. His late-in-life chance-of-a-lifetime had come after he had invented a rainmaking machine that he said, matter-of-factly, might "bring peace to the Middle East." Salter envisioned the machine rising more than a hundred feet above the ocean, its rotating curved blades drawing power from the wind as it sprayed water up. It was shaped like a giant eggbeater blade, improbably spinning up out of the waves.

The idea was to humidify the air to create rain clouds off the coasts of dry areas. Salter calculated that each of the machines would deliver skyward roughly 2 million gallons of water an hour, and that several hundred such devices could deliver to the Middle East the equivalent of "the flow of the River Jordan" in annual rain. "Almost everybody in the meteorological world said there's no way it would work," he'd acknowledge years later. In fact, computer modeling based on atmospheric physics showed that the device would dry up the coasts, rather than moisten them, since the droplets it would add to one area would lead to the reduction of moisture elsewhere. "Nobody expected that the effect would be so powerful in the wrong direction," he wrote.

A scientist in Edinburgh heard about Salter's idea and introduced him to a British cloud physicist named John Latham. Latham explored clouds typically on paper, occasionally in small airplanes, and he had been inducted into the Royal Society for having established in a precise way how lightning formed. Latham had an idea of how to counteract global warming by brightening clouds. But he didn't know how to produce the droplets of seawater that his scheme required.

Salter, a few years his junior, was his man. At ages seventy-two and sixty-five, the pair of Blue Teamers dreamed wildly. Soon after meeting, they'd articulated in phone calls and e-mails their shared vision: to brighten Earth's ocean clouds through the hourly addition of a trillion trillion droplets of seawater, about a billion tons a year. The salt from the particles would make the water in the clouds form more droplets. As a result, the clouds became shinier, reflecting more radiation back into space. The oceans would steadily cool. Sucking carbon dioxide with chemical engineering or algae blooms changes the makeup of the atmosphere. Brightening clouds, like the Pinatubo Option, would reduce the amount of solar energy that struck the world.

Salter worked seven days a week as an unpaid professor at the University of Edinburgh to invent a boat that would, as he would say, save the world. He considered utilizing wave power devices tethered to the seafloor to run the sprayers, but Latham told him that the sprayers would have to move so as to brighten clouds in different areas. A boat with solid surfaces known as wing sails didn't quite work. So he borrowed an idea invented eighty years before called a Flettner rotor. It was a type of nautical propulsion system that used neither sails nor lines but odd masts with fins on them that provided thrust perpendicular to the wind.

The ship that Salter invented looked like a floating toboggan topped with three spinning chimneys shaped like corkscrews. Instead of smoke, the chimneys would release a fine mist of seawater. Latham was delighted. Together they envisioned a fleet of fifteen hundred such vessels, 300 tons each, brightening clouds just enough to reverse the warming caused by our carbon sins. Unmanned, Salter's boats would meander Earth's oceans and hack its clouds, ghost ships following a GPS spell cast by geoengineers.

Frustration had shaped both men's lives, as had ingenuity and luck. Salter recalls his early days as an apprentice helping build the SR177, a British jet fighter he says proudly "could set the altitude record in the morning and the speed record in the afternoon." The British Ministry of Aviation scrapped the project in 1957, and Salter believed (as did others) that the decision was improperly influenced by the aerospace industry's concerns. Salter went on to construct an early hovercraft and a robot named Freddy, designed to interact with a five-year-old. ("When the university realized how complex the mind of a five-year-old child was, they ended the project," he said.)

As an engineering professor at the University of Edinburgh in 1972, Salter invented a 300-ton floating canister, shaped like the top half of an exclamation point, which generated electrical power from ocean waves. It bobbed in the water, so they called it Salter's Duck; and like Freddy, it sits in the National Museum of Scotland. But despite the recognition, in 1983 the engineer found himself in a war with the British government's energy agency over funding for wave power.

The bureaucrats preferred nuclear energy over renewables, and his grants ran out. "They were changing numbers, withholding reports, changing failure rates," he said, and he provided testimony saying as much to a House of Lords committee convened to look into the issue. He felt unappreciated and often bitter.

Latham went from humble roots at Imperial College London to become a prominent cloud scientist at the University of Manchester. Many years ago he was walking with his son Mike near a cottage outside the Scottish village of Waunfawr, a thousand feet up, slightly higher than rows of still, shimmering clouds out over the Irish Sea. In the distance, just out of view, sat Ireland. Gazing out over the expanse, his boy looked straight ahead at the sunset and asked, without looking at his father, "Why are those clouds so shiny?" Latham explained that the clouds were wet, and that the water droplets in the clouds were reflecting sunlight in all directions.

"Like mirrors," said Mike. "Soggy mirrors."

A few years later, on a narrow Scottish road in 1986, a soldier driving in the wrong lane hit Latham's car head-on. Miraculously, Latham and his two passengers, both children, survived the accident. But a doctor examining Latham later found in his head a tumor, which was immediately removed. Shaken by the turn of events, and already inclined to leave the bureaucracy of the university, Latham began to look for work that could afford him more freedom. A new institute to study Earth's climate had opened north of London, so Latham, to get work, set out to write some papers on climate, a subject he knew little about.

He stumbled on a scientific paper by a scientist named Anthony Slingo that suggested that increasing the number of low clouds on Earth by 20 percent could block enough sunlight to offset the warming that a doubling of CO_2 would cause. (Scientists call them "stratocumulus" clouds; they are found over more than a third of the surface of the ocean.) The paper also suggested that simply making the existing clouds brighter could have the same effect. Making drops in clouds one fifth smaller would do the trick. (A cloud with smaller drops and the same amount of water has more surface area from which to reflect light.) The British atmospheric scientist was close friends with Bernie Vonnegut, the novelist's brother and the inventor of a widely used rainmaking technique. Perhaps, Latham thought, there might be a technique of brightening the clouds manually.

"Soggy mirrors," he recalled his son saying.

Pollution from ships brightened clouds in their wake, as visible by satellite. Could this be done on purpose? Latham imagined utilizing the salt particles found in the sea itself to change the clouds' own properties. He calculated that to do that, and to compensate for the warming caused by carbon dioxide, would require lofting roughly twenty-two pounds of the ocean's own salt into clouds above every square mile of ocean. "Control of Global Warming?" was how the *Nature* editors titled the short letter, published in 1990, in which Latham explained his idea. "A bit of fun, maybe it could work," said Slingo, who worked at the institute, and he hired Latham soon after. The letter accumulated a bemused, negligible reaction and dissolved into the archives. And then, thirteen years later, scientists started to become quite alarmed over global warming, and engineer Salter met scientist Latham.

Soon after meeting, the pair began publishing papers on their cloud-brightening concept, by themselves or with colleagues. A computer modeling study suggested the scheme might work on a global scale. If it didn't, or if there were worrisome consequences, they'd shut the boats down, argued Salter; ten days later, the oceans would be as before. That might be a key advantage over the Pinatubo Option, Latham said. Another possible advantage was that the boats might allow cloud brightening in targeted areas, perhaps near the Arctic, to protect polar ice. Scientists have disagreed over whether the Pinatubo Option could be deployed regionally.

Still, the pair couldn't get a penny to pay for their research into the boats. In 2005 a government bureaucrat showed up at the University of Edinburgh, and they explained the entire scheme to him, proposing a modest research program. A month later, the response from the Department for Environment, Food and Rural Affairs came back: no. On his Web site, Salter included responses he had prepared with Latham to the rejection letter point by point.

Department: "Not yet soundly proven."

Latham/Salter: "Three papers on [cloud brightness] have been published. . . . The predictions have been confirmed."

Department: "We are not at the stage where there is a strong push for radical alternatives."

Latham/Salter: "Are you so confident in your present policy that you were sure that there is no need for any backup plan?" ("They mismanaged foot and mouth outbreak, they mismanaged mad cow disease," Salter fumed later. "All they did about climate change was to go around the world saying how awful it's going to be.")

And so with few options left, the pair agreed to cooperate on a one-hour broadcast on the Discovery Channel. The show's conceit involved a studly British physicist, a female green innovator, and an acerbic male eco-entrepreneur. During each episode the three put outlandish climate solutions through the paces. For the taping, Salter and Latham flew to Cape Town, South Africa, where conditions might be suitable for brightening clouds. A Florida boat builder managed to convert a tri-catamaran into a Flettner boat, which sailed like the ghost ship Salter had designed, though it didn't spray.

Instead of producing salt particles from seawater, the producers hired a pyrotechnics company called Big Bang Stunts and Effects, and had them rig three hundred flares full of tiny salt particles to a small boat floating offshore. On camera, the test flare they tried didn't rise high enough to have an impact, the physicist reported. "In the battle with global warming, the team is losing ground," intoned the narrator. But the crew fired the flares anyway, and whether by skill or divine intervention, a small white cloud appeared amid the smoke. On the show, which aired during the summer of 2008, Salter gave a thumbs-up, and calculated that the cloud had blocked enough sunlight to negate the carbon emissions of "several power stations."

But months later he was less sanguine. "It would be a chance in a thousand that would work again," he said, calling it a fluke result achieved under unrealistic conditions. "The data were mostly useless."

Six months later, Salter and Latham had their hats out in front of the richest man in the world. It was within the stone-and-wood halls of the University of Edinburgh where they would make their case. The meeting, Salter told me beforehand, "is where we decide whether someone's going to give us money. There's a bunch of guys from the Gates Foundation coming and they're coming along to decide whether we're good enough." (The other reason for the meeting was to develop joint research work.) Salter called it "the big event," and he had prepared special slides, labeled "Gates," to make the pitch.

It was a motley crew of about two dozen at the meeting: researchers both clean-cut and ragged, a few bureaucrats, a neatly dressed tech executive from Silicon Valley named Kelly Wanser, and a handful of reporters, including me. One of the scientists had turned up at breakfast still buckling his belt.

While the crowd included no employees of the Gates Foundation, the cast of characters included a few with close ties to the billionaire. Two scientists were in charge, ostensibly—Ken Caldeira and David Keith, who each year quietly distributed the $1.5 million that Gates gave them for geoengineering research. Sitting in the back was a smiley and mysterious executive from Microsoft named Karen Fries, a close colleague of Melinda Gates. Was it Fries's job to report back to Redmond on whether the ghost ships deserved to be funded? Or would Caldeira and Keith decide? They wouldn't say, and no one quite knew, but the promise of Gates's money hung in the air like water vapor. (The tacit understanding was that scientists were not to mention his name.)

Despite the impressive credentials of the assembled scientists, and the distances from which they had traveled for the meeting, there was a marked informality to the affair. The words "[Type the company address]" appeared on one of the distributed handouts. As Wanser opened the meeting, Salter stood unexpectedly—the schedule didn't have him speaking for another five hours. He pointed at a recorded video playing of undulating low-lying clouds that was slowly hypnotizing us. "What I want you to do is to see how wildly turbulent clouds actually are," lectured the mechanical engineer. Some of Britain's finest cloud physicists looked on with quizzical expressions. "What you're really seeing here are like a bunch of rollers under a moving object," Salter remarked. His "big event" was off to a rocky start.

Engineering mock-up posters of Salter's cloud-brightening wind-powered-toboggan-boats hung along the side of the classroom where we sat. Human figures depicted in rather creepy outline, with little wrenches, stood here and there in the drawings, providing an illustration of scale. They drove home the still abstract idea that people might try to brighten ocean clouds someday soon. And yet clutching their tools, the figures seemed competent, ready to adjust the spraying equipment or call Salter on their cell phones.

Before the digitally conjured little men sat real little men, imagining how to purposefully alter the wide ocean. The day's morning talks would focus on the atmospheric science behind the scheme, with the afternoon devoted to spray technology. (Out of roughly six hours of sessions, twenty minutes were set aside for "nonscience considerations" such as geopolitics or ethics.)

Latham's introductory presentation reviewed the basics of the idea. To compensate for global warming, the technique would have to create liquid drops at a rate of 30 metric tons of seawater per second across the world. UNACCEPTABLE RAMIFICATIONS MIGHT BE DISCOVERED, one slide noted. Latham quoted from his 1990 paper, asking whether the idea could "inhibit or neutralize global warming." He looked up with old, wide eyes. "Twenty years later the question is still open," he said. Salter nodded.

That question had three parts. Could brightening clouds cool the planet, and would there be side effects? Second, could scientists actually brighten clouds, using droplets of seawater? And finally, could geoengineers actually make the droplets properly at sea?

Presenting first was an American scientist named Phil Rasch, who discussed modeling simulations performed on a supercomputer to address the first question. The verdict: the scheme would provide cooling, "to the first approximation," as one researcher put it, if the salt particles could be successfully delivered into the clouds. ("A big if," said Latham.) The computer suggested that the effects would be drastic, and seemed to depend on where the process was attempted. Seeding a quarter of the world's seas, including the Indian Ocean, off the Horn of Africa, and large swaths of the Pacific would counteract the warming that a doubling of CO_2 worldwide would cause. "We get substantial replenishment of the sea ice," said Rasch.

Scientists worried that reflecting more of the Sun's rays into space instead of allowing them to strike the ocean would rob the system of energy that produced rain, the same concern scientists have with the Pinatubo Option. Dry spots included swaths of ocean off South America and other areas. Also, marine cloud brightening cooled the oceans but didn't change the temperature of land—another similarity it shared with the Pinatubo Option. That could worsen monsoons, which derive energy from the difference between the two. There also was worry that the technique would create cold conditions across the Pacific that could disrupt rainfall patterns, especially in South America. Fries looked concerned.

What ensued was a comparison of encouraging results from unreliable models compared with the risks that the same unreliable models suggested. One scientist, for example, said that debating how the scheme might affect rainfall was a waste of time, since computer models simply couldn't predict precipitation. But Salter, from a seat near the front of the room, seemed confident in climate modelers' abilities. "I think what we've got is a model of a musical instrument and what Phil has done is learned to tune it," he said with a hint of peevishness. "I'm speaking with all the knowledge of a mechanical engineer—we don't know how to ride a bicycle when we start, but we end up doing it." (With the same air of confidence, months later, Salter would suggest that the ghost boats could be deployed if necessary "to apply a bit of damping" to "get a controlled amount of La Niña [a Pacific seasonal rain pattern]." But a climate scientist said that doing that would result in "overwatering some regions and underwatering others . . . a real disaster.")

Following the modelers came the cloud scientists, whose world consists of water molecules and the specks around which they condense and form clouds. The ratio is roughly a million million water drops to every speck, be it salt, aerosol droplet, or biological bit. Atmospheric physicists such as Latham call these tiny, crucial particles cloud condensation nuclei; their presence in the sky provides a surface onto which chaotic wet air takes form, creating clouds. Just how aerosol particles in the sky form and alter clouds is one of the biggest mysteries in climate science, which makes the cloud brightening scheme all the more audacious.

Getting droplets from Salter's ghost ships up into the clouds may not be so easy, the cloud scientists said. The trick would be to raise the seawater mist from the sprayers, which lofted them upward a few dozen yards at most, to the sea clouds, which sit more than 2,500 feet above the ocean surface. At nighttime, in a process called convection, moist, warm air from near the sea surface rises to provide water droplets for clouds. That would help the cloud-brightening boats. In the daytime the Sun heats both the ocean and the atmosphere, so without the temperature difference there's a possibility that mixing would halt. One of the scientists estimated that 1 percent of the particles that Salter's boats created would get to the clouds. "My guess would be 3 percent," remarked Salter, whose frustration was getting harder to mask.

Assuming the salt particles managed to arrive in the clouds, would they actually make the droplets of the clouds smaller, as Latham and Salter were banking on? On average, marine clouds have 30 to 200 particles per cubic meter. The sprayers would have to quadruple that or more to have the most potent effect. A pair of German scientists had written that that was next to impossible, since the new cloud droplets would coagulate and form larger ones, causing rain. In Edinburgh, scientists wondered whether the new seed particles could interfere with the drops that create clouds naturally. In addition, smaller drops tend to evaporate faster than big drops, said Rasch. That would mean that if brighter clouds turned out to be more fleeting ones, they might need to brighten many more. In this case, Salter's boats would actually make clouds less shiny in aggregate than doing nothing. Experiments at sea to sort it all out were the answer, the atmosphere experts said. Salter seemed irritated; a simple demonstration with a boat or two and some satellite imagery would be enough, he objected.

But in the last session of the morning, David Keith questioned whether a boat was needed at all. He urged the group to think about the fundamental design questions—how much energy per droplet of seawater it made sense to expend, for example—instead of getting too caught up on particular technologies. "There's no point pounding one of the nails down into the table," he said. Might airplanes deliver particles to brighten sea clouds more efficiently than boats? "In some e-mails I had a couple years ago with John Latham we talked about some Soviet jet-powered water bombers," Keith told the group, which soon headed off to lunch.

Salter was seething as he hurried back to the classroom after the meal. He had hoped the day's meeting would focus on a technical evaluation of his boats and their spraying systems. That would allow them to focus on reliably producing droplets smaller than thirty millionths of an inch wide out at sea, a challenging task, or filtering the water, which could clog the spray nozzles. This talk about airplanes, he said, was maddening. Latham said 30 tons of salt water across the whole world would be required to be lofted every second for his scheme to work. For Keith's idea, Salter said, you'd need to utilize cargo jets such as the C-130, which has a capacity of less than 20 tons. "If he is doing it by airplane you've got to have a plane taking off every second," he fumed. "What an irritating distraction."

Salter arrived at the classroom. There two university technicians had arranged a pair of apparatuses he had designed that released mist into the air, illustrating the sort of tiny drops the technique required. The one shaped like a bong was coolly breathing vapor; the cup-size one had stopped. "I'm not quite sure why," an assistant told Salter. "It looked good earlier, and then it got pretty anemic," Salter said, fiddling with the device. Salter looked at the assistant. "They'll all come back in a cluster, so when you see them coming, really fire it up," he said.

As the crowd shuffled in for the afternoon session, Salter and Karen Fries happened to bump into each other as they were exiting the room. Salter stepped aside, signaling grandly. "After you," he said, smiling. She blushed.

Salter's session began with some basic discussion of reflectivity and the principles behind his boat concept, and he went on to describe the flows of water and energy within his ghost ships, delving into detailed engineering drawings. "Speaking for myself, I'd be keen to hear some numbers about what energy costs, focusing on the spray," said Keith, who along with Caldeira was the scientist Salter had to convince to earn the Gates funding. Salter paused and obliged, mentioning the energy requirements for pushing water up the mast: "We have an airstream here that's going to use 12 watts"—Caldeira interrupted him to ask why freshwater was needed, and minutes later he asked about the specifications of his nozzle.

"Will you let me just show you the next slide," snapped Salter. He went on about the filtration system and pressure flows within the machine and the device's flow regime and a valve he hadn't "designed yet" and parts he calls eyelids and "gun-drilled holes that are bringing the freshwater up to this region."

"It would be helpful to hear more of the design trade-offs," Keith said as Salter manipulated the mouse. "This is a detailed design drawing."

"I am trying to get the energy consumption on this part of the process as low as possible," Salter said.

"Why as low as possible?"

"Because I think energy in the middle of the ocean is expensive. Now, you may disagree with that. You're going to have a Hercules C-130 taking off every fifteen seconds; you're not worried about energy consumption," said Salter, his voice rising.

Keith betrayed no emotion. "So is there a number, is there a target, is it 'low as possible' "?

Salter explained that the goal was to minimize the amount of equipment the boat would require. But Keith pressed on. "From my point at this stage, the detailed design drawings aren't the issue," he said. Salter argued that without money for research, designing equipment was all he had been able to do. "We've taken design quite a long way. Now, if we hadn't done that, people could say, 'Oh, well, there's no chance of getting this filtration working in the middle of the sea,' " said Salter. He went on to describe an experiment he proposed to try brightening clouds off the Faroe Islands, north of Scotland, and a Scottish colleague described a technique they had developed to make microscopic nozzle holes by cutting holes in silicon wafers. The wafers would serve as a nozzle with tiny holes to make jets. By vibrating the seawater before spraying it the jets would form droplets. But they need money to actually build and test a unit. The scientists estimated that to succeed, the clouds would have to be seeded with drops smaller than thirty millionths of an inch across—using only wind energy at sea. Few techniques to do that existed, apparently.

Then came Armand Neukermans, a gentlemanly Flemish physicist who had helped develop the first inkjet printer while at Hewlett-Packard. He had a different approach than Salter, and was ostensibly vying for the same money. Neukermans began with gracious praise of his hosts. "I find this a beautiful concept, essentially," he said of Latham's idea. An evisceration of his competitor followed, as he performed the best exposition on advanced droplet-making anyone of us could imagine being delivered. The spray-technology extravaganza was a sight to behold: antique graphs photocopied from old papers, spectroscopy curves printed with labels printed by a dot matrix printer in all caps, handwritten illustrations scanned right onto the slides, patent #3990797, the advantages of spherical spray heads.

Salter had alienated the very scientists he needed to impress. Neukermans impressed them. Neukermans needed dry erase pens. Keith jumped up to get them for him. Neukermans described how air blown in a perpendicular direction would spread the drops apart, though the term the engineers used, he said, was "pissing in the wind." Fries laughed brightly. By applying an electrical current to the nozzle, Neukermans explained, the charged drops would repel one another, preventing them from touching each other and getting too big. The bottom line was that the Flemish expert thought he could make the droplets the right size. It appeared he just needed some money to get started.

"You may think I have a bias against silicon," Neukermans said at one point. "Actually, I made a substantial fortune [with it]," he said, inadvertently applying a fine mist of salt water into the Scottish inventor's wounds.

Kelly Wanser, whose Silicon Valley can-do spirit had made possible the day's meeting, led the final session of the day. She and Salter were both builders, both determined, both business oriented among rather academic colleagues with little experience turning ideas into real-world equipment. When I first met her she had click-clacked through the lobby of a hotel in Washington, D.C., in long black boots and stylish but unfriendly glasses. Five sips into her Sapphire and tonic she'd used the phrases "ad hoc," "VCs," and "wire-frame plan." In a field dominated by disheveled, aging men, she probably provided needed professionalism, I thought. (Geoscience + engineering) × Wanser = Geoengineering.

Her presentation split tasks into "a technical development plan," a "scientific plan," and what she called a "policy and communications activity." The progression of research was to flow from "desk scale" to "point scale" to "1/1,000," and then to "global." "Advise—does it make sense to proceed to the next phase?" one slide read. That would be a small-scale ocean trial.

But Ken Caldeira seemed concerned with the pace of the conversation. "We underestimate the interest that NGOs will have in these activities and the possible blowback from ill-considered reaction of governments," he said. "Too early field testing done by this group could end up shutting down a lot of activities that might be valuable." Salter just frowned, looking through some of the handouts. (Wanser told me later that the experiments the scientists envisioned attempting would fall under "existing environmental regulations," though she said some experiments would be small enough that they wouldn't require "an EPA permit." That said, she believed it was important that whatever work they did be conducted in an "open, scientifically rigorous" manner.)

The discussion moved on to unfair attitudes toward geoengineering research; the group agreed that scientists pursuing the field lacked the recognition they deserved. The issue seems particularly relevant to Salter and Latham, given their advanced age. Standing to wrap up, the moment seemed to crack Wanser's professional exterior a little. "I want to say my last word is to . . ." said Wanser, quivering for a second as tears appeared in her eyes. "I didn't mean to get emotional, not appropriate," she said, righting herself. "In terms of what the message is about why we're doing this is, I think John and Steve"—another pause, as more tears formed, scarcely visible behind her glasses. "Here you have two people who over the course of a decade have created a vision of—if we ever have to do geoengineering, let's do it well." There was a smattering of applause and I felt inspired, a little weirdly, to clap, too.

The group ended up that evening at a restaurant called Amber, lodged in an ancient stone building beside the regal medieval castle that looms above the Edinburgh town center. A palpable optimism was in the air as the wine flowed and the table got noisy. Wanser had ditched her CEO glasses and was wearing her hair down, gamely listening as her neighbor explained the finer subtleties of the government's management of satellite systems. Latham and Caldeira calculated how a spray system might be used to cool the city of Abu Dhabi by 1°C or more. But Salter seemed sullen, keeping mostly to himself. I asked him if he was satisfied with his presentation. "I don't think some people liked it," he said with a frown.

As the meal wound down, David Keith stood up. "I want to give a toast to Stephen and John for being our hosts, and for coming up with this idea," he said, raising a glass. "It's a great idea and it takes real persistence to push an idea for a long time." Salter smiled slightly, nodding. He felt the Gates money slipping from his fingers. A few minutes later, having donned his coat, he faced Karen Fries by the front of the restaurant as the group was filing out. He implored her to support his work. She looked at him patiently and tried to be gracious. Salter looked right at her. "We are running out of time," he said.

"If I need to hire someone to make me a spray nozzle, I think I would hire Armand, and not Salter," Ken Caldeira told me later. And, sure enough, Neukermans emerged with some of the Gates money to develop the spray idea, roping in willing engineers both young and senior to help him develop sprayers. ("The meeting was a disaster for Steve," John Latham told me.) Kelly Wanser, meanwhile, started a nonprofit they dubbed the Silver Lining Project to help coordinate the work. It included dozens of scientists volunteering their time, possible tests in a German cloud chamber, and a proposed small-scale ocean experiment involving American and British scientists, to study the effects of the technique on clouds. Half a year after the Edinburgh meeting, Neukermans was showing initial success making a single stream of droplets of the right size. But each boat would need billions of streams, said Salter. "I'm not sure his method would work. It might. I'm not arrogant enough to think that I'm the only one with the right answer," he said.

Meanwhile, Salter worked on a new concept while he waited for his next chance. The idea was to use wave power to diffuse hurricanes at sea. Some of the thinking for the idea had come together at a meeting with other inventors in Seattle, at a well-known inventions company called Intellectual Ventures, in which Gates was an investor. The proposed device, which looked like a giant floating Brita filter with its top cut off, worked by pumping millions of gallons per hour about 600 feet down from the surface. The machine was a plastic tube that employed a series of valves that would allow water into the tubes but not out of them, sending 150 kilograms of water per second down. Removing the warm water from the surface would cool the surface and possibly save "New York from the next Katrina," Salter wrote in a prospectus on the idea. When a U.S. patent on the idea appeared, "William H. Gates III" was named as one of the coinventors with Salter, ironically.

The Scotsman had some hope of funding: half a year after the Edinburgh meeting the British Royal Society published a report that recommended the U.K. government spend £10 million on geoengineering research, though government agencies didn't seem particularly enthusiastic about it. He quoted from a famous Rudyard Kipling poem, *If*: "If you can meet with Triumph and Disaster/And treat those two impostors just the same." "It's one my favorites," he told me. "My wife hates it, though."

HUNDREDS OF SPECIES OF INSECTS HAVE BEEN SUCCESSFULLY utilized as so-called biological control agents, deployed as an alternative to pesticides. But scientists often don't understand why some control agents work and others don't. And when they don't, as the bizarre case of the spotted knapweed illustrates, things can go bizarrely bad. Spindly with purple flowers, the weed is among the invasive scourges of the American West, covering 4 million acres of land in Montana alone. Native to Asia and Europe, it poisons other plants and crowds out important species in a variety of Western ecosystems. In the 1970s scientists went to war against the plant by importing a natural enemy, the gall fly. The fly makes the plant's seed head its nest by laying eggs in special sacs the plant creates to defend itself, sapping energy the plant needs to make seeds, and thereby weakening its ability to spread.

But research published in 2006 by the U.S. Forest Service showed that three decades after the gall fly was deployed, the strategy had failed to control the knapweed plant. What happened instead is that deer mice learned to climb up the plants' stalks so they could eat the larvae. As a result, instead of largely dying out over the winter when their food supply is covered in snow, the mice populations have skyrocketed.

And with them, so has disease. In areas with introduced gall fly populations the Forest Service found three times the amount of mice carrying hantavirus compared to areas without the fly. Hantavirus is a disease that spreads through urine and droppings. There's no evidence that any humans have gotten the virus, which can lead to kidney damage or respiratory failure as a result, but scientists are worried that it might happen. "It illustrates the complexity of how these things play out in the system," Forest Service scientist Dean Pearson, who published the work, told the *New York Times*. "The chain goes all the way to humans." Experts say that other biological techniques using insects to control plant populations could also backfire. The hantavirus discovery, says Svata Louda at the University of Nebraska at Lincoln, "is the tip of the iceberg. . . . We don't know what we're doing when we mess up natural systems."

The Right Side of the Issue

"Meet me in ten minutes in the courtyard outside the EPA offices, by the sculpture of a rose," David Schnare tells me.

It's springtime 2008, a bright day, but there's something decidedly noir about this rendezvous. I've never spoken to him before. He knows I'm a reporter at *Science.* I know he has a Ph.D. in environmental science and works by day at the Environmental Protection Agency. He moonlights as a Republican attorney at the Thomas Jefferson Institute for Public Policy, a right-leaning advocacy group based in Virginia. The year before, in little-noticed testimony before a Senate committee, he stated that geoengineering "prevents more damage than exclusive reliance on carbon control," which raised eyebrows. He was the Blue Team's first public voice on Capitol Hill.

On a public e-mail group devoted to geoengineering Schnare had recently asked the rest of the Geoclique to sign on to a letter calling for geoengineering research. He said he would send it to the National Academy of Science and, cryptically, "certain others." To me that meant the government, so I'm following a hunch that he is lobbying Capitol Hill to fund geoengineering research. A week before, I'd gotten through to a Senate budget aide, but he denied that they were considering the idea. Then I called Schnare.

I see a man with a mustache standing by a bench. He looks a bit agitated. "Sit down," he says. "I know you called over to the Senate." I feel my pulse quicken as I stare at him, his nostrils flaring. "There won't be any funding for geoengineering research, and you're the one who did it," he says. I blink.

Apparently the legwork I had done as a reporter had gotten back to him. But my hunch had been correct, apparently. Or at least it was likely to have been correct. He wouldn't be upset if he hadn't been in contact with the Senate aides, I figure. "Why didn't you call me first?" he says. The occasional office worker wanders by.

Behind closed doors, Schnare tells me, he had managed to convince staffers on one of the Senate's appropriations committees to consider paying for geoengineering research. They'd just silently tuck roughly $5 million into the bowels of an enormous spending package they were preparing, he envisioned, labeling it something vague, innocuous. He hoped the money would be used for "planning, laying out research plans, figuring out what we need to do" to study planethacking in the federal government, he tells me, "everyone together, in one room, to brainstorm what kind of research to do." He glares at me. "Now there's no chance of that."

The Senate staff, he explains, knew how controversial the notion of government-funded geoengineering was. So when a reporter came calling, they'd denied it and figured Schnare had leaked news of the proposal. As a result of my call, Schnare claims, big money for geoengineering studies is off the table for 2009.

This is about the "future of the planet," as Schnare puts it, and that future is in jeopardy. I don't think planethacking was a good idea, I say, but I tell Schnare that I think scientists should examine it. So I feel a combination of guilt and incredulity over the turn of events. (I'd never manage to confirm what fraction of the story was true.) I ask him what he thinks about the risks of global warming. "Look, I don't drink the IPCC Kool-Aid on climate change," he declares. (The UN climate science panel would win the Nobel Peace Prize for its work on climate science later that year.) "But if the warming is happening, I'm telling you, whatever the cause, we have to be ready in case things get really bad."

• • •

Why do some of the same people who believe human activities are not warming the globe—or that climate change isn't a crisis—feel that geoengineering is required to fix the problem?

The essential issue at the center of the global warming debate is how much impact, essentially, humans have on the planet. If one fears the worst-case scenarios—say, Al Gore's version of the world—then one will tend to believe that humanity has had an outsized role in shaping Earth's recent history, through relatively small amounts of greenhouse gases emissions that are wreaking increasing havoc. Liberals have tended to align with most climate scientists on this question, believing that humanity has messed up the planet by pouring carbon dioxide into the atmosphere. Their solution, then, has been to endeavor to change human behavior and stop pouring it in.

If instead one thinks that the planet actually doesn't much react to human influences, one's likely to believe that carbon dioxide won't have much of an effect on climate and that rising levels of it aren't much of a problem. American conservatives, in large part, have generally gravitated to this position. They discount the worst-case scenarios, viewing the planet as largely resistant—on a global scale, at least—to what physicist Alvin Weinberg in 1967 called the "taints" of modern technological life. For the most part they have viewed carbon dioxide as a harmless gas—or even beneficial in higher quantities.

But by the mid- to late 2000s, as the poles melted and the climate warmed, the conservative position was becoming untenable, and the position among deniers of global warming science was in crisis. Some had argued for decades that global warming was not happening, or as Oklahoma senator James Inhofe had put it, that it was a giant "hoax." But the facts about global warming had become all but inescapable. Al Gore won a Nobel Peace Prize in 2007; even Exxon-Mobil acknowledged the science of man-made climate change. (Minor scandals in late 2009 related to emails among climate scientists and a handful of errors in the IPCC's 2007 report did little to weaken the case for anthropogenic warming.)

What happened was that conservatives adapted their positions to reflect the increasingly clear reality that man-made carbon dioxide is having a significant effect on the planet. Schnare is not the only Republican who questions the science of climate change and who nonetheless allows for the possibility of catastrophe. In a 2009 column, George Will of the *Washington Post* first complained that "alarmists" were unnecessarily concerned about the warming, given the recent flat trend in yearly temperature averages. He went on, however, to cite a report that warned of a "cataclysmic warming increase . . . *even if nations fulfill their most ambitious pledges concerning reduction of carbon emissions*"—his italics—as though to suggest that the impact of carbon dioxide in the atmosphere was so great that it was too late to do anything about the problem.

Will's confusion is, of course, understandable. So far it appears that the planet's temperature could be very sensitive to greenhouse gases, which is why scientists are so worried. But we don't know *how* sensitive. That uncertainty is an inescapable part of the climate conundrum, but it has particularly confounded conservatives who would wish to avoid tough measures to cut carbon dioxide pollution. "Seizing upon either the low end of the projected rise (to argue for complacency) or the high end (to argue for fatalism) is a silly exercise that utterly fails to comprehend probability and statistical range," wrote *New Republic*'s Jonathan Chait, critiquing Will's position.

While the impacts of climate change have become more apparent and increasingly worrisome, conservatives have sought a solution that could allow them to address worst-case climate scenarios while clinging to their core beliefs about humanity, the world, and the proper place of government in people's lives. Geoengineering is steadily becoming their essential tool to do so. Like a climate policy Swiss Army knife, it has proven useful to support a number of talking points on the subject. First, the promise of geoengineering as a technical fix to the problem has allowed conservatives to present a solution to global warming instead of being seen as simply blocking liberals' proposed carbon regulations. To do so, for example, economist

Bryan Caplan of George Mason University calls geoengineering "the best option we have" to address global warming, given its cost. Block the Sun but continue to spew billions of tons of carbon dioxide into the atmosphere? "There's a good chance it could be that easy," he says. The apparent ease at which the Pinatubo Option might allow influence on the climate gives conservative geoengineering advocates the opportunity to make misleading comparisons with climate policies that would slow human greenhouse emissions.

Strategies that involve blocking the Sun turn a pollution problem—there's too much carbon dioxide in the air—into a temperature problem—it's too hot. That fits with a longtime argument among climate denialists that global temperature rise primarily results from solar activity or natural cycles, and not carbon dioxide. By championing a technique that directly alters the temperature of the planet instead of the composition of the atmosphere, conservative advocates of geoengineering have a "solution" that fits the argument they been making all along. Schnare has even argued that geoengineering offers a "middle ground" for the climate debate because it allows for an emergency response but makes unnecessary "high-cost strategies" such as efficient buildings or capturing carbon from coal plants. And conservatives are increasingly citing liberal distrust of planethacking as evidence that they don't really want to solve the problem—or even that they have more ulterior motives.

His attempt to convince Congress to pay for planethacking may have fallen short, but Schnare's efforts were only the tip of an organized and influential iceberg. The spring of 2008 would see geoengineering emerge as a new focus for the right wing of the climate policy crowd. In June of that year, the American Enterprise Institute, Washington's premier right-wing think tank, embraced the push for geoengineering research with the first of six planned workshops on the topic. Figures from the right wing of U.S. politics have become fixtures at the regular meetings on planethacking. The week after Schnare and I met, for example, Deputy Secretary of Defense Paul Wolfowitz and leading nuclear hawk Fred Ikle of the Center for Strategic and International Studies were part of an invitation-only discussion of geoengineering hosted by the Council on Foreign Relations. And the following year, Danish statistician Bjorn Lomborg, author of the best-selling *Skeptical Environmentalist*, weighed in with a report on geoengineering from the controversial Copenhagen Consensus Center. An outspoken opponent of the Kyoto/Copenhagen process, Lomborg called a number of geoengineering options "promising responses to global warming" while "carbon taxes and cap-and-trade policies are very poor answers."

Though more contrarian than politically partisan, the 2009 book *SuperFreakonomics: Global Cooling, Patriotic Prostitutes, and Why Suicide Bombers Should Buy Life Insurance* provided new fuel to the growing yes-to-geoengineering-no-to-emissions-cuts position. In a chapter devoted to geoengineering research, authors Stephen Dubner and Steven Levitt acknowledged a risk that "the greenhouse gases we've already emitted *do* produce an ecological disaster." But launching a global crash program to lower carbon emissions is a "costly, complicated" solution, they wrote. Transforming the energy system to reduce carbon emissions? Shouldn't be "dismissed," they wrote. But they quoted approvingly from scientists who called wind power "cute" and solar power "probably not" a good solution—and they basically ignored the potential of nuclear power to provide carbon-free power. Compared to lowering emissions, they wrote, the Pinatubo Option, is a "fiendishly simple plan." And conservatives like Bret Stephens of the *Wall Street Journal* opinion page applauded the best seller. He called it "delightful."

There is no single "conservative" position on humanity's proper role in nature, of course. But various right-wing opinions on the topic might each be considered one of a variety of what might be called a muscular set of views. Theodore Roosevelt, for example, was a hunter as well as a naturalist and believed in safeguarding nature for its own sake—but also as a means to enrich citizens' lives. So while he set up dozens of nature preserves and signed environmental laws, he also created the federal Reclamation Service, which eventually transformed the nation by creating millions of acres of farmland through dams and irrigation. Philosopher and conservative icon Ayn Rand, writing in 1957's *Atlas Shrugged*, described the central purpose of men's lives as "remaking the earth in the image of one's values." While Jimmy Carter asserted that environmentalism demanded a recognition that "our great nation has its recognized limits," Ronald Reagan drew a sharp contrast and was able to successfully tap American thirst for technological progress in response. "Conservative environmentalism believes environment protection is a good like any other, i.e., a thing that one rationally trades off against other goods to obtain" as opposed to a "moral good," wrote Robert Locke in *FrontPage Magazine*. Right-wing blogger Dave Nalle says that conservatives, perhaps naturally, are conservationists, but in a way that promotes "improving the Earth" through active means—say, growing trees—as opposed to erasing humanity's footprint.

It was certainly through active means that prominent Blue Teamer physicist Edward Teller proposed the Pinatubo Option in an op-ed published in the *Wall Street Journal* in 1997. It promoted "contemporary technology" to geoengineer as a "more realistic" option than emissions cuts. "Let's play to our uniquely American strengths in innovation and technology," he wrote. After George W. Bush was elected in 2000, Teller wrote a letter to the new administration urging it to launch a research effort in geoengineering.

Less than a year after Bush took office, a bureaucrat at the Department of Energy named Ehsan Khan, an acolyte of Teller's, took an interest in the controversial idea. Khan was a former Pentagon official navigating a geeky bureaucracy of solar panel buffs and former academics. Geoengineering wasn't his first foray to the edge of the scientific mainstream. He had arranged meetings in 1998 with scientists pursuing "zero point energy"—a source of power supposedly found in quantum voids. He would go on to serve as the Department of Energy's point man on the notorious hafnium bomb project—an ill-fated effort to make a golf ball-size nuke. (Critics said the concept broke the laws of physics.)

In the fall of 2001, Khan arranged a seven-hour meeting by teleconference with government scientists in Washington and Albuquerque to brainstorm a geoengineering research program. The meeting was one of a series meant to craft the Bush climate research effort. "There is a significant risk of rapid and disruptive climate change in the decades ahead," said the thirty-seven-page draft report Khan subsequently produced. Among the hazards it described were "rapid climate change in the Arctic" and "super hurricanes." To potentially "avert the severe consequences through deliberate actions" Khan's write-up proposed $64 million in research funds for modeling, engineering, and field tests on a variety of geoengineering schemes. Carbon emissions cuts were not mentioned.

One might have thought that Bush would be the perfect president to introduce geoengineering into the American political lexicon. His administration sought technological solutions to a variety of complex sociopolitical problems. He pursued the scientifically dubious concept of a space-based missile shield while shunning several arms-control treaties; to fight the wars in Iraq and Afghanistan, the Bush administration spent billions on high-tech weaponry such as unmanned drones while failing to establish a sufficient cadre of translators with knowledge of the languages spoken in those countries. On global warming, too, Bush's government cast the challenge as a long-term one, and purely technological. It was not "command and control regulations" that would solve environmental problems, Bush said in 2003, but "technology and innovation," while Dick Cheney denigrated conserving energy in one's home as a mere "sign of personal virtue." Bush committed $1.1 billion to an international consortium building a fusion reactor in France and expanded the government's energy research, including a $1.2 billion initiative to build hydrogen cars.

But Khan faced roadblocks at every turn when he tried to add geoengineering research to the emerging Bush climate science program. For six years, until the end of the Bush administration, his report never saw the light of day, despite Khan's behind-the-scenes efforts. Some Department of Energy scientists believed the reason was that the draft report laid out dire risks—a message in stark contrast to the White House line on climate change, which was to deny that the risk of such disasters could be quantified. "The perception was that [geoengineering research] would make it appear as though we didn't have any confidence in the [stated] technological approach," says Jerry Elwood, then an official at the Department of Energy's climate change program. "One of my jobs was to protect the secretary of energy from adverse publicity, regardless of the merits of the research," explains former top energy department manager Ari Patrinos, now with a California biotech firm. He actively stymied the Khan report, he said. "If the Bush administration was seen promoting this kind of work, there would be the usual 'There they go again.'" Compounding the problem with the message was the medium: Khan's interest in far-out ideas made him "just not in the mainstream here at DOE," another bureaucrat in the department explained.

By focusing the conversation about climate change on geoengineering, conservatives have a new way to recast the climate problem that takes carbon dioxide completely out of the picture. Keeping the Pinatubo Option as a worst-case scenario in their back pocket allows them to appear to act responsibly while avoiding its cause: the greenhouse effect, which humanity has exacerbated by burning fossil fuels.

It can be difficult to remain scientifically credible while questioning the link between man-made carbon dioxide and global warming. Dubner and Levitt faced intense criticism from a variety of researchers in large part because *SuperFreakonomics* raised doubts about a variety of aspects of settled climate science, most notably the atmospheric role of carbon dioxide. That gas, they assert, does not "necessarily" warm Earth, nor have warming trends in Earth's past followed rises in the carbon content of the atmosphere. "So hopelessly wrong," was how climate modeler and numerical analyst

William Connolley described the book before listing ten errors or obfuscations he found. Having responded directly to few scientific critiques, Dubner and Levitt characterized their attackers as "ideological" opponents or, tellingly, "carbon crazies." On a radio program, Steven Levitt put it plainly. "The real problem isn't that there's too much carbon in the air. The real problem is it's too hot." It's easy to understand the appeal of the Pinatubo Option for those armed with the belief that the climate problem can be divorced from the carbon one. "Geoengineering is intended to be a large-scale response to climate change, whether human or natural," wrote David Schnare on an online discussion group in 2009, citing "significant argument as to the causes of global warming."

Others have offered geoengineering as an alternative option to emissions cuts after arguing that the latter won't solve the problem. In 2009 Alan Carlin, an EPA economist, cited a "significant" debate over the usefulness of emissions cuts when he became the center of a brouhaha. The fight was over an internal analysis of climate science he wrote about proposed regulations on greenhouse gases. Carlin believes that the concept that human activity is warming the planet is a "hypothesis," and in his analysis he argued that certain temperature records suggested it was "unlikely" that greenhouse gases "have much effect on measured surface temperatures." He went on to question whether the EPA should accept conclusions of the Intergovernmental Panel on Climate Change. Conservative critics pointed to leaked e-mails in which Carlin's boss told him not to intervene on the issue, calling them an example of scientific censorship. (The EPA said that Carlin's comments had been properly analyzed and that he could only publish them on his own Web site, which he did.) And in interviews with reporters, Carlin suggested that the Pinatubo Option was preferable to cutting emissions because it was cheaper and because, unlike reducing greenhouse gases, "it would actually work."

Carlin told me in 2008 that geoengineering could not only protect us from warming but also, depending on what techniques were used, would allow humanity to prevent "the next ice age" in case it comes, an argument Edward Teller had made.

But climate scientists have dismissed this line of thinking as unscientific or worse. After Schnare suggested to the online discussion group that geoengineering could be useful regardless of the cause of warming, scientists there summarily rebuffed him, including David Keith, who nineteen minutes after Schnare's message wrote, "There are people who passionately believe there are aliens on Air Force bases in Nevada, and likewise there are folks who have very strong opinions about how the climate science is fundamentally wrong." NASA climate scientist Gavin Schmidt called Carlin's work "a ragbag collection of un-peer-reviewed Web pages, an unhealthy dose of sunstroke, a dash of astrology."

If for some on the anti-Kyoto right geoengineering has allowed them to cling to the idea that global warming is unrelated to carbon, for others discussing the concept is like waving an enchanted fairy wand. It magically transforms them from a nasty climate science–denying Gargamel into a friendly and enlightened Smurf who practically quotes from *An Inconvenient Truth*.

Take, for example, key officials at the American Enterprise Institute (AEI), which has arguably done as much to prevent the U.S. government from passing limits on greenhouse gas emissions as any U.S. congressman, K Street lobbyist, or AM radio talk show host. For right-wing activists that respected institution has served as an intellectual bridge between activists and prominent and powerful figures in Washington. Beneath the banner of inclusivity, its well-apportioned meeting rooms in Washington have hosted various prominent deniers of climate change, including the late novelist Michael Crichton, who called mainstream climate science "shockingly flawed and unsubstantiated" at a 2005 lunchtime speech there.

Visitors to the AEI's global warming page on their Web site on June 3, 2008, for example, would have read that the idea that "the science is settled" was not "generally true" when it comes to the global warming debate. But that day happened to be an event featuring presentations on geoengineering. The director of AEI's project to study geoengineering, Sam Thernstrom, exemplified the new conservative respect for the consensus climate science in his opening remarks when he cited research findings from none other than the conservatives' bête noire, James Hansen. The NASA scientist had long been derided by the right as an "alarmist" for his dire warnings. But Thernstrom cited Hansen's belief that greenhouse gases in the atmosphere could lead to "irreversible catastrophic effects." This risk, said Thernstrom, was among the reasons to justify geoengineering research.

The idea that geoengineering is available as an option has been used in conjunction with the belief that cutting emissions won't—or can't—fix the problem and, not surprisingly, that geoengineering represents a better solution. To make that case, conservatives have exaggerated the known ease and effectiveness of geoengineering while dismissing the idea of lowering carbon emissions as being expensive and impractical.

In discussing the relative effectiveness of cooling the planet manually and reducing the warming we are causing with greenhouse gases, climate deniers regularly fall into the trap of oversimplifying its usefulness. "Geoengineering would provide more time for the world's economy to grow while investors and entrepreneurs develop and deploy new carbon-neutral energy sources to replace fossil fuels," wrote Ronald Bailey of the magazine *Reason.* "With regulations and rations, you command five billion people to change. On the other hand, with solar radiation management, you toss Earth a beach umbrella and get on with your life," wrote conservative columnist Neil Reynolds of the *Toronto Globe and Mail.* Teller, for his part, wrote in his 1997 *Wall Street Journal* op-ed piece that geoengineering "is not a new concept and certainly not a complex one." The AEI's Thernstrom states that cooling the planet using the Pinatubo Option offers "three powerful virtues in a climate policy that mitigation, at the moment, cannot claim." They were, he said, "fast," "affordable," and "effective."

The problem with the argument that geoengineering can be used to forestall emissions cuts is that every ton of carbon dioxide that gets emitted into the air while we're delaying could increase the future calamity. Even Tom Wigley of the National Center of Atmospheric Research, as Blue as a member of the Blue Team gets, believes that geoengineering must be used simultaneously with emissions cuts, as soon as possible. The longer we wait to geoengineer, the more severe the geoengineering we might need in the future—hence the greater chance of side effects. Plus there's the problem—mentioned in chapter 4—that deploying the Pinatubo Option while continuing to pour carbon into the atmosphere could spell disaster were the geoengineering to stop for some reason. Scientists who have studied radical geoengineering approaches hardly consider them "simple" or, as Thernstrom calls the Pinatubo Option, "effective."

We just don't know what their side effects would be or whether they would behave the way we think they do. Physicist David Keith has his own triplet to describe sun-blocking techniques such as the Pinatubo Option: "fast," "cheap," and "messy." It's one thing to say that geoengineering techniques appear as though they would cool the planet and so they should be studied. It's another to say they're good enough to be considered an alternative to getting to the root of the problem, too much carbon dioxide.

The extent to which they're "affordable" is also up for debate. The early estimates are certainly relatively cheap. In 1997, for example, Teller estimated that the Pinatubo Option would cost "between 0.1 and 1.0% of the hundred billion dollars a year that is estimated would be required to price-ration fossil fuel usage" to reduce emissions. Intellectual Ventures, the company profiled in *SuperFreakonomics*, puts the cost of deploying the Pinatubo Option globally at $150 million to start, with a yearly operating cost of $100 million.

The 2009 report on geoengineering that Lomborg published included a detailed economic analysis of the issue. Its authors were the AEI's Lee Lane, and J. Eric Bickel, of the University of Texas at

Austin. The pair modified a well-known economic model designed to study the relationships among greenhouse gas emissions, economic growth, and damage from climate change. They analyzed three approaches: Salter's cloud-whitening boats, the Pinatubo Option, and launching giant orbiting sunshades into space. For each, they tried to analyze potential benefits in terms of trillions of dollars in economic value of avoided climate impacts. To do so, they translated changes that geoengineering might cause to radiative forcing—the direct energy the Sun transmits to Earth—into changes in temperature.

The pair estimated costs in terms of possible direct side effects, such as the Pinatubo Option's impact on rainfall, though for indirect costs—for example, how geoengineering schemes would affect agriculture—the pair said that "the literature offers virtually no guidance." The results: cost-benefit ratios of roughly 1 to 25 for the Pinatubo Option and roughly 1 to 5,000 for the cloud-whitening strategy. (They dismissed the orbiting sunshade.) A panel of five top economists whom Lomborg organized, including three Nobel Prize winners, went on to rank cloud whitening as its number one "solution" for climate policy, putting carbon emissions cuts way down on the list, at number twelve. (It sat below "technology transfers" and "expand and protect forests," though a carbon tax was number two.) The work "makes it clear that" cutting emissions with the Kyoto-Copenhagen approach is a strategy "we need to rethink," said Lomborg in a press release.

In a companion paper, however, policy analyst Roger Pielke Jr. called the numbers in Lane and Bickel's paper "at best, arbitrary, and more critically, not grounded in a realistic set of assumptions about how the global earth system actually works." The vast uncertainties plaguing the analysis, he said, allowed the authors to obtain numbers that reflected the authors' biases. And the modifications that the authors used to change the economic model were crude, said Pielke. For the Pinatubo Option, for example, the changes simply assumed that adding aerosols to the stratosphere would change the direct energy the Sun was transmitting to Earth, the so-called radiative forcing.

But putting haze in the atmosphere could alter temperature in much more complex ways, for example by changing patterns in ocean currents, rain, or seasonal patterns such as El Niño. Others questioned the whole exercise of ranking "solutions" given the pervasive uncertainties. "How can you vote on which solutions are most cost-effective if you don't even know if they work?" wondered environmental scientist Alvia Gaskill, a prominent member of the Geoclique.

Conservatives have used the reservations that scientists and climate activists have with the radical idea of geoengineering as proof of nefarious aims. The argument goes that if you are uncomfortable with planethacking but support tackling the climate crisis with emissions cuts you are seeking political control, money, or vast political change. In 2008, when the AEI announced it would be holding a workshop on geoengineering, David Hawkins of the Natural Resources Defense Council expressed qualms online, worrying that the AEI "has aligned itself with the [climate] denialist camp." For such temerity, David Schnare called him an "archetypal environmental advocate" in an angry rebuttal. Hawkins's preference for mitigation, Schnare claimed, revealed the environmentalist's long-standing bias toward measures that would harm low-income consumers. "David [Hawkins] never met a wealth transfer [from poor to rich] he didn't like," he said. (After the exchange, Ken Caldeira established rules on the discussion board prohibiting personal attacks.)

The *Toronto Globe and Mail*'s Reynolds wrote approvingly that the Pinatubo Option "would require no changes in lifestyles, no sacrifice in standards of living . . . perhaps this helps explain why it is neither discussed nor researched by environmentalists or governments. Environmentalists almost always select solutions that require changes in lifestyle and standards of living. Governments almost always select solutions that expand bureaucracies." (Given the stakes, one wonders if a global program to alter the temperature of the planet might involve significant government infrastructure to implement, oversee, and regulate—not to mention manage the inevitable international squabbles over the thermostat.) In 2009, writing on the *National Review*'s Web site, Jonah Goldberg linked approvingly to an article about Lomborg's study, writing, "make significant progress on global warming at minimal cost without declaring war on capitalism? Crazy!"

The belief that liberal environmentalists could never support geoengineering research led David Schnare to declare that advocates for planethacking research "have no friends" in the Obama administration. But liberals *are* calling for geoengineering research, including environmental pioneer Stewart Brand; ecologist Tom Lovejoy; climate advocate Rafe Pomerance; and Lord Nicholas Stern, author of the authoritative 2006 Stern Review, which called for emissions cuts. Clean Air-Cool Planet and the Clean Air Task Force, nonprofits based in Portsmouth, New Hampshire, and Boston, respectively, quietly cosponsored a June 2009 briefing on geoengineering in Washington, D.C. Notably, no major environmental groups have opposed research into geoengineering. Mostly they have avoided saying much. And that's because they don't want to embolden climate change deniers who might seek to use the idea of geoengineering to undermine the effort to stop dumping carbon dioxide into the atmosphere.

The Geoclique's Blue Team does, in fact, have friends in the Obama administration. Secretary of Energy Steve Chu, for example, has said that painting exterior surfaces such as roofs white to reflect sunlight could be a positive step to combat global warming—a technique that would reflect sunlight and is unrelated to greenhouse gas controls. His science deputy Steve Koonin led a study on the Pinatubo Option in the summer of 2008 before joining the Department of Energy. Obama science adviser John Holdren helped organize the Cambridge meeting on geoengineering mentioned in this book's first chapter.

Schnare declared geoengineering's lack of supporters in the Obama administration in the spring of 2009. Holdren had just scrambled to clarify his position on geoengineering after an Associated Press reporter quoted him saying that "we have to look at" geoengineering solutions, given the climate risks. Four months into his job at the White House, Holdren told the AP reporter that he had raised the issue of geoengineering "in administration discussions" on climate policy, though only as a "last resort." A few hours later, the AP reported that "Tinkering with Earth's climate to chill runaway global warming . . . is being discussed by the White House as a potential emergency option." Holdren then responded angrily, saying in e-mails to administration officials and outsiders that he was "dismayed" that the reporter had twisted his words to suggest that the White House was, as implied, "giving serious consideration to geoengineering." "I may never talk to the press again," wrote Holdren to administration colleagues. (Months later, unease with the topic was evident within science agencies like NASA, following the lead of the White House.) Holdren was concerned about exploring geoengineering at a time when the new administration was pushing Congress and its international partners to embrace emissions cuts. The hesitation among some on the left, including those in power, not to publicly embrace geoengineering research is a strategic choice, not an ideological one.

There's real basis for the concern that the concept of planethacking could be misused for political ends. Former National Coal Board scientist Richard Courtney is among Britain's biggest opponents of emissions cuts. Like many on the right, he calls the idea of man-made global warming the Anthropogenic Global Warming hypothesis, or AGW. "I am firmly convinced that dangerous AGW is not a problem and cannot become one. However, I do think the possibility of geoengineering should be supported. My reason for this is a political ploy," wrote Courtney in 2009. "The politicians need a viable reason if they are to back off from this commitment to [carbon] constraints without losing face. The geoengineering option provides the needed viable reason to do nothing about AGW now."

In April 2009, Michael Totty, an editor at the *Wall Street Journal*, asked Ken Caldeira over a series of e-mails to write a "strong advocacy piece" about geoengineering for a special pullout section of the newspaper on the environment. "As much as I'd like to have an essay that makes a forceful case for adopting an immediate program of geoengineering, it seems that no credible expert is willing to go that far," Totty wrote. Caldeira was wary, so Totty suggested that Caldeira write that "all the attention on mitigating the cause of global warming is causing us to ignore a crucial strategy" and that "we're going to have to take this whole different approach." Totty called it "Geoengineering Now."

"I have problems with the implication that we need to take a 'different' approach, rather than 'an additional approach,'" Caldeira wrote back to the editor. He declined the assignment.

Two months later the piece was published, written by a writer, not a prominent scientist. Caldeira and Keith, along with others, sat in a hotel lobby perusing the article, whose headline was "It's Time to Cool the Planet." A large illustration showed Earth sitting on ice with fans cooling it off. The moderate stance of the article suggested that Totty had been unable to convince the writer to take the aggressive position he had wanted.

"Their job is to sell newspapers," said Keith.

"And to advance the right-wing agenda," said Caldeira, laughing. "It is Rupert Murdoch."

For years, advocates of geoengineering research have feared that as the concept of geoengineering becomes better known, just talking about the concept will dull momentum for regulations on carbon emissions. If the perception existed that an easier solution was available, would governments grow complacent? A related question was whether conservatives would make efforts to use geoengineering to convince them to do so. I asked Keith if this particular fear was coming to pass. "So far, it's been a red herring," he told me. But others are not sure. "Delay is the Carbon Lobby's strategy," wrote Alex Steffen, in an essay called "Geoengineering and the New Climate Denialism" in 2009. The idea of geoengineering, he wrote, has become "part of the new attempt to stall [emissions] reductions."

IN THE 1980S, NONGOVERNMENTAL ORGANIZATIONS TACKLED the problem of the desert encroaching on northwestern Kenya by turning to a hardy savior: the thorny mesquite tree. Indigenous to the Americas, the tree sustains dry environments across the world. It provides shade for people and animals. Its leaves fertilize the ground, and its ten-foot-long roots reach for nutrients. Mesquite trees reduce erosion and can lower the salinity of soils in which they grow.

But three decades since their introduction into Kenya, mesquites have become a nightmare, invading grazing fields and croplands. In 2009 the Kenya Forestry Research Institute reported that the tree possibly was responsible for the demise of local populations of the acacia tree, which had declined by 40 percent in some regions. Farmers said their animals lost their teeth after chewing on the tree's tough wood. The government estimated that 66 million acres of the land were at risk from the spread of the species, which farmers call the "Devil Tree." Mesquite invasions have also ravaged areas of Australia, the United States, and Ethiopia. Government officials in Kenya, as well as in Ethiopia, have urged farmers to adapt to the spreading menace. "Pastoralists must now start making use of the benefits that come with the species instead of looking at the negative aspects of it all the time," said a spokesman for the Kenyan Forestry Service, which has looked into using the tree's pods for food and its wood for charcoal and as material to export as flooring.

But farmers, having lost land on which to graze their animals, are hesitant. "If there's a use for this tree, only Allah knows it," a farmer named Ibrahim Hamadou told *The Independent* (London) in 2004. Hamadou "walks on average 25 miles each day in search of land free of the Devil Tree on which to graze his animals," the paper wrote.

A Political Climate

The nations of the world fuss and argue over trade, air pollution, monetary policy, overfishing, human rights, Palestinian refugees, Third World debt relief, immigration, the Armenian genocide, slavery, greenhouse gas emissions, and monetary policy. And plenty else. They fight with guns over religion, communism, capitalism, borders, oil, terrorism, and more.

So why should one believe they'll cooperate over geoengineering?

A looming threat that nations faced in the twentieth century was other nations. Scientists developed a technology to address the problem: nuclear weapons. And then the solution became its own problem. During the Cold War the world only narrowly averted a full-scale nuclear exchange; the legacy of the Manhattan Project is the risk of a nuclear-armed Taliban.

A looming threat nations face in the twenty-first century is catastrophic climate change. Scientists are developing a technology to address the problem: geoengineering. Is this proposed solution to become its own problem? It was on my mind on a spring day in 2009 as I explored the verdant grounds of the Calouste Gulbenkian Foundation in downtown Lisbon. A two-day, closed-door event on geoengineering and geopolitics would commence there the following day, the first-ever international meeting on the topic.

Calouste, you got us into this mess, you miserly kook, I thought. Or guys like you. The foundation's benefactor was an Armenian oil baron who had built his $280 million fortune off Iraqi crude in the early decades of the twentieth century. I couldn't help wondering whether this famous oddball—a paranoid hypochondriac, serial renter of limousines, and connoisseur of teenage mistresses—would have felt guilt over the atmospheric sins his greed was to engender after his death in 1955. If so, the meeting was part of his penance. The statutes of his noble foundation declared "charitable, artistic, educational, and scientific" purposes. Under that aegis the experts here would meet to imagine the international havoc geoengineering might wreak and how to avoid it.

Scheduled from noon to noon to accommodate the Continental business schedule, the two-day confab brought high-level European bureaucrats, Red and Blue team scientists, and a few reporters together in a conference room adorned with bottled water. The meeting had been organized by the International Risk Governance Council, which in the past had tackled topics such as nanotechnology and genetically modified crops. But those issues were relatively noncontroversial compared to geoengineering. Techniques such as cloud whitening or the Pinatubo Option offered geopolitical challenges like almost no other technology. Few international rules covered their potential use. They were relatively cheap and therefore could be deployed by a wide variety of countries, acting alone or in concert. "It is not implausible that in a few decades the option of geoengineering will look less ugly for some countries than unchecked changes in the climate," said an article in *Foreign Policy* that predicted a number of ways by which international conflict could arise as a result of the technology.

Carnegie Mellon University engineer Granger Morgan employed a smiling cartoon Sun on his PowerPoint presentation to illustrate the various techniques to the attendees. Then he explained the dire geopolitical implications of their potential use. The Pinatubo Option or cloud whitening might include winners and losers, such as some countries that got more rain and others that got less. Shielding Earth from the Sun wouldn't prevent carbon dioxide from continuing to acidify the ocean. That might doom others who depend on the ocean's harvest, which would be threatened by the loss of crucial ecosystems. The Pinatubo Option might damage the ozone layer, or cause global temperatures to skyrocket if it was halted in the future. "Once you get into it, you're sort of committed," Morgan said. "So given all of these uncertainties my initial reaction was that we ought to just create a taboo against geoengineering just as we have created a taboo against chemical and biological weapons." But then, he said, he realized humanity might need it.

"I can envisage this subject becoming one of the most contentious and polarizing issues in the climate change debate unless it is addressed now," said the council's Donald Johnston. He said he hoped the meeting would set in motion "the global debate" to begin to come to terms with the new technology. Nations had certainly failed to confront international technological risks in the past. "After World War II Oppenheimer tried to set up international control of the uranium supply; we really should have done that," said John Steinbruner of Johns Hopkins University. "It was a big mistake not to have tried."

A number of the good-government types from Europe who attended the Lisbon event had hardly heard of geoengineering before the confab. ("I've spent my life trying to undo the effects of technology—now I'm here," an official from the European Environment Agency lamented at one point.) But the members of the Geoclique who helped organize the Lisbon event were hoping to lay the groundwork for field tests of the controversial idea. The scientists wanted to move beyond paper studies of geoengineering without getting their research efforts banned, and do it in a way that minimized the chance of unexpected damage to the biosphere. Few expected that the concept would be an easy sell to the public, but it was crucial to get started. The *Foreign Policy* article, like most of the Lisbon attendees, called for "a serious international research effort." That was one thing that was needed, it said, and the other was "developing norms to govern and manage the risks" of planethacking. "One of the disaster scenarios is that we get to a panic situation, and there are pressures to deploy something that might work, but in fact had we done five or ten years of straightforward research we would know [if] it's a really dumb idea," said Robert Lempert, a scientist with the RAND Corporation.

Conducting that research meant confronting a daunting catch-22. It will be hard for scientists to accurately describe the risks of various planethacking approaches with any level of certainty without conducting small-scale tests. So even if they performed the tests, lots of uncertainties would remain about the dangers—including the side effects that scientists knew about and those that were unknown. Opponents of field tests would point at the same uncertainties, however, as a reason to stop the experiments—even in the early stages.

Depending on the technology, scientists often find themselves either resisting international regulation or, alternatively, presenting to the public a regulatory framework they had set up themselves to provide society the sense that their work is under control. (Scientists working with biowarfare agents have sought to create their own rules, and it's mostly been successful.) In the case of geoengineering, at least judging from this small group, scientists had opted for the latter strategy, and fairly desperately wanted to establish international guidelines to regulate field tests. "The obvious question is how do we develop some transparency and some oversight, while at the same time not making things so ponderous in terms of international approval that we can't get anything started," said Morgan. There wasn't much time—small-scale experiments into geoengineering were coming fast, David Keith said. "We need to start thinking now about what norms and institutions ought to govern these experiments," said Keith.

Depending on the scientists, the participants in the meeting either understood the depth of their public relations challenge or sorely underestimated it. In any event, there was a kind of desperation to the scientists' desire that the public accept field tests. Recent history was not encouraging in this regard. LohaFex, which was deploying its experiments in arguably the most desolate place on the planet, had nearly been shut down despite little or no risk to marine life. More audacious plans seemed destined to run up against strenuous opposition. "We're talking about putting a material in the air that will cool the planet directly—that's a really difficult message to manage," the council's Chris Bunting told me over breakfast. A set of public focus groups convened in 2009 for the U.K. Royal Society found that participants were "generally cautious, or even hostile, to geoengineering." A poll of a thousand British adults over age sixteen found that 47 percent of respondents thought the Pinatubo Option should not be considered for use, more than twice as many of those who said it should. A Royal Society staff member, Andrew Parker, said that he felt "we need to convince people" that field testing for geoengineering was acceptable.

"If in fact we get to the point that we suddenly discover we've got a climate emergency, public perceptions are going to change very rapidly," Morgan said. Given the grave situations under which planethacking would be deployed, participants at the meeting suggested that whole tenets of environmental thought might require new interpretation when considering geoengineering field testing. Environmental lawyers often invoked the "precautionary principle," a loosely defined term that generally refers to the idea that an action shouldn't be taken if science suggests there is a *possibility* of harm. Under the precautionary principle, considering an action with uncertain consequences generally puts the burden of proof on those who might create the risk. (It's a relatively new idea in environmental policy, though it is an active principle in European law.) The great risks of climate change itself, said Belgian government science advisor Jean-Pierre Contzen, could allow proponents of geoengineering to argue that it was "precautionary" to deploy large-scale solutions—even though those techniques themselves had lots of uncertainties. "Instead of considering the precautionary principle an obstacle to geoengineering, if properly used it could be an ally," he said.

But in the years before the most dire impacts of climate change came to pass, this particular argument might not hold much sway with policymakers or the nonprofit organizations that influenced them. Recent experience with iron fertilization experiments was an early test case. Several of the scientists commiserated during a break over what they viewed as a particularly frustrating episode surrounding the technique: a 2008 resolution passed by nations party to the 1992 Convention on Biological Diversity. That was the ruling that split Germany's government over LohaFex in 2009 and restricted ocean fertilization experiments to "small-scale research studies within coastal waters." (See chapter 8.)

Though the convention's statement wasn't binding law, the public sentiment against ocean geoengineering experiments that it helped encourage was the subject of a presentation in Lisbon by Richard Lampitt. He was a jocular oceanographer who took off a scuffed leather jacket to deliver his talk. He had contributed special robotic collection devices for the LohaFex iron fertilization experiment months before the Lisbon meeting, and clung to e-mails from the ship as its fate was debated. In the press, he said, the idea had taken root that that the ocean was "a pristine environment" that should not be manipulated "to counteract our own overconsumption of resources." Public outcry about the mission had been severe, and "the scientific community became too relaxed" to dispute various misconceptions about the technique, Lampitt said. "Fish will all die" from low oxygen levels the experiment would cause was one misconception, he said. "The press, media latched onto these very adverse, very negative messages," Lampitt continued, and as a result "the whole thing nearly fell apart." His conclusion: "Public anxiety should not be ignored" as a factor determining policy. In other words, I thought, the scientists and officials in the Lisbon conference room wouldn't be the ones to make the rules.

Other international bodies apart from the Diversity convention gave scientists some hope that governments would listen to them. Margaret Leinen, formerly of Climos, spoke more positively about a pair of international agreements called the London Convention and the London Protocol, which regulate the dumping of waste or chemicals on the high seas. With close consultation from scientists, both agreements had effectively modified their declarations to cover iron fertilization without fully banning it, despite the fact that the agreements had never been designed to regulate scientific research. But the scientists said they weren't out of the woods yet. Under way was a process to define how nations would certify proposed fertilization experiments as "legitimate scientific research." Toward that end, for example, Greenpeace had proposed, among other requirements, that any scientist wishing to do the experiments had to obtain "prior informed consent" by all eighty-six nations party to the London Convention. Such rules would make it "very difficult for the scientific community to get approval to do any research," said Lampitt. "I do want to be responsive to public concerns but I do want them to be sensible concerns."

Deciding just what was sensible would depend heavily on the conditions of the experiments. Without some sort of international rules for experiments in place, feared Ken Caldeira, even very-small-scale and demonstrably benign field tests could spark a backlash that could shut down other efforts. He mentioned the Latham-Salter scheme to brighten clouds, of which he expected publicly announced field tests "in the next few years." That highlighted the idea that the public would come to its own conclusions in regard to the intent of field experiments. Intent, he suggested, might loom large as a concept on the international stage. "It's entirely clear that the experiment itself is benign at the scale of a single sprayer but it's also equally obvious that it's the first step down a slippery slope," he said. "Now we as scientists could say if I was testing the sprayer just because I'm designing misters to do cooling for a ship or some other application there would be no question about it. But when it's the first step toward a geoengineering thing, it introduces the notion of a slippery slope." (Some time later, tongue firmly in cheek, Caldeira would remark that a recent decision of his to run his air conditioner with the windows in his car down, an obviously futile effort to cool the atmosphere, was an example of intentionally "geoengineering the planet for thirty seconds.")

Caldeira knew firsthand how damaging public controversy could be to even small-scale experiments planned for natural habitats. He had helped organize an international experiment to inject more than 20 tons of liquid carbon dioxide onto the floor of the ocean on Hawaii's Kona coast in 2001. Biologists had shown that the effect on local ecosystems would be minor at best and the amount of carbon dioxide added to the water minuscule compared to the amount of the gas belched out by Hawaii's undersea volcanoes. The scientists were top-notch—from respected Department of Energy labs and MIT, among others. And 20 tons of carbon dioxide? The ocean holds as much as 38 trillion tons of carbon annually, the scientists argued, so what they were adding was a minute amount.

But in 1999, before the scientists could implement a public outreach strategy to hear out local environmentalists and fishermen, an article titled "Feds to Test Impact of Dumping CO_2 into Kona Waters" appeared in *West Hawaii Today*. A local coalition of opposed environmentalists formed, some of whom affixed bumper stickers to their cars labeled "Stop CO_2 Dumping." One letter published in *West Hawaii Today* compared the scientists to "Hitler's doctors and scientists," telling them to "go home and poison your own water."

Caldeira found a number of leaders of the opposition reasonable, technically proficient, but most of all, politically shrewd. "They knew the experiment was benign, but they saw it as the leading edge of a wedge and they understood that if you wanted to stop ocean carbon sequestration you stop experimentation," he said. After hundreds of letters in opposition flooded the Department of Energy and the offices of state lawmakers and officials, the project was postponed and eventually died. A similar experiment planned for the Norwegian Sea was canceled after similar intervention by environmental groups. Subsequent experiments on a smaller scale than the planned Kona one showed that carbon dioxide released underwater "did exactly what the MIT guys said it would do," says

33 *Since the data about* William D. Nordhaus, "An Analysis of the Dismal Theorem," Cowles Foundation Discussion Paper No. 1686, January 2009.

34 *"The built-in pipeline inertias"* Martin Weitzman, "Reactions to the Nordhaus Critique,"draft document, February 7, 2009.

34 *"The less clear the science is"* Michael Tobis, "Uncertainty Does Not Call for Inaction," Only in it for the Gold blog, http://initforthegold.blogspot.com/2008/06/uncertainty-does-not-call-for-inaction.html (June 10, 2008).

34 *"uncertainty regarding global warming"* Thomas Schelling, "Uncertainty and Action on Climate Change," Project Syndicate, www.project-syndicate.org/commentary/schelling1 (January 2008).

35 *"All of this"* Martin Weitzman, "On Modeling and Interpreting the Economics of Catastrophic Climate Change," *Review of Economics and Statistics* 71 (2009): 1.

---

37 *started an experiment* José Cattanio et al., "Unexpected Results of a Pilot Throughfall Exclusion Experiment on Soil Emissions of CO_2, CH_4, N_2O, and NO in Eastern Amazonia," *Biology and Fertility of Soils* 36 (2004): 102–108.

---

3. The Point of No Return

39 *"Tipping points, once considered"* Richard A. Kerr, "Climate Tipping Points Come in from the Cold," *Science* 319 (2008): 153.

39 *"If a very small warming"* Ibid.

40 *"We've created a climate change scenario"* Peter Schwartz and Doug Randall, "An Abrupt Climate Change Scenario and Its Implications for United States National Security," report by Global Business Network to the U.S. Department of Defense, October 2003.

40 *"Ecosystems as diverse as"* *United Nations Environment Program 2009 Yearbook*, www.unep.org/geo/yearbook/yb2009/PDF/3-Climate_Change_%20UNEP_YearBook_09_low.pdf , 21.

40 *Human activities are responsible* Tim Appenzeller, "The Case of the Missing Carbon," National Geographic Online, http://ngm.nationalgeographic.com/ngm/0402/feature5/ (February 2004).

41 *In 2008, scientists published* Tim M. Lenton, Hans Schellnhuber, et al., "Tipping Elements in the Earth's Climate System," *Proceedings of the National Academy of* Sciences 105 (2008): 1786–1793.

41 *as little as a 4°F rise* David Adam, "Amazon Could Shrink by 85% due to Climate Change, Scientists Say," *Guardian*, March 11, 2009.

41 *Reviewing data one evening* Andrew C. Revkin, "In Greenland, Ice and Instability," *New York Times*, January 8, 2008, Science section.

42 *Steffen couldn't believe* Konrad Steffen and Jason Box, "Surface Climatology of the Greenland Ice Sheet: Greenland Climate Network 1995–1999," *Journal of Geophysical Research* 106 (2001): 33,951–33,964.

42 *It has lost nearly half* David Carlson, "International Polar Year Present and Future Changes in the Arctic," *International Polar Year*, prepared congressional testimony, 2009.

42 *That's why the Arctic* David Biello, "Global Warming Reverses Long-Term Arctic Cooling," *ScientificAmerican.com*, September 4, 2009, www.scientificamerican.com/article.cfm?id=global-warming-reverses-arctic-cooling.

43 *In 2007 scientists with the Intergovernmental Panel* *Climate Change 2007*, the Intergovernmental Panel on Climate Change (IPCC) Fourth Assessment Report, Contribution of Working Group I, ed. S. Solomon, D. Qin, M. Manning, Z. Chen, M. Marquis, K. B. Averyt, M. Tignor, and H. L. Miller (Cambridge, U.K.: Cambridge University Press, 2007), "Summary for Policymakers," fig. SPM.3, 13.

43 *"We can't really afford to wait"* Anil Ananthaswamy, "Sea Level Rise: It's Worse Than We Thought," *New Scientist* (July 1, 2009).

43 *11 feet of sea level rise* Jonathan Bamber et al., "Reassessment of the Potential Sea-Level Rise from a Collapse of the West Antarctic Ice Sheet," *Science* 324 (2009): 901.

44 *the Larsen B ice shelf collapsed* Ananthaswamy, "Sea Level Rise."

44 *five smaller shelves on the peninsula* Ibid.

44 *The mighty Jakobshavn glacier* Krishna Ramanujan, "Fastest Glacier in Greenland Doubles Speed," NASA Feature (December 1, 2004), www.nasa.gov/vision/earth/lookingatearth/jakobshavn.html.

44 *Holland discovered fisheries data* David M. Holland et al., "Acceleration of Jakobshavn Isbrae Triggered by Warm Subsurface Ocean Waters," *Nature Geoscience* 1 (2008): 659–664.

44 *"That warmth now has"* Richard A. Kerr, "Winds, Not Just Global Warming, Eating Away at the Ice Sheets," *Science* 322 (2008): 33.

45 *"only during the summer months,"* Michael MacCracken, "On the Possible Use of Geoengineering to Moderate Specific Climate Change Impacts," *Environmental Research Letters* 4 (2009).

45 *"Holding off on geoengineering,"* Michael MacCracken, e-mail to geoengineering Google group, September 17, 2009, www.mail-archive.com/geoengineering@googlegroups.com/msg02463.html.

46 destroy *sea ice using special ships* Ibid.

46 *Scientists also have proposed* Douglas R. MacAyeal, "Preventing a Collapse of the West Antarctic Ice Sheet: Civil Engineering on a Continental Scale," *Annals of Glaciology* 4 (1984): 302.

47 *Corn yields in Italy and France* David S. Battisti and Rosamond L. Naylor, "Historical Warnings of Future Food Insecurity with Unprecedented Seasonal Heat," *Science* 323 (2009): 240.

47 *four megadroughts that rocked* Edward R. Cook et al., "Hydrological Variability and Change," Synthesis and Assessment Product 3.4, U.S. Climate Change Science Program, 144, www.climatescience.gov/Library/sap/sap3-4/final-report/.

47 *They decimated the U.S. West* Ibid.

47 *A federal report published in 2008* Ibid., 210.

47 *disquieting to consider* Ibid.

48 *It's a crucial issue for farmers* Battisti and Naylor, "Historical Warnings," 241.

49 *Methane lasts roughly* "Methane," Environmental Protection Agency, http://epa.gov/methane/.

49 *As the methane supercharged the greenhouse effect* Fred Pearce, *With Speed and Violence: Why Scientists Fear Tipping Points in Climate Change* (Boston: Beacon Press, 2008), 90.

49 *Ocean circulation patterns shifted,* Ibid., 104, 148.

49 *"Life on Earth was transformed"* Pearce, *With Speed and Violence,* 92.

51 *enormous glacial lake of freshwater* W. S. Broecker and R. Kunzig, *Fixing Climate: What Past Climate Changes Reveal about the Current Threat—and How to Counter It* (New York: Hill & Wang, 2009), 111.

51 *Temperatures abruptly plummeted* Pearce, *With Speed and Violence,* 149.

51 *"perhaps even a single season"* Ibid., 150.

51 *a "drunk [on] a rampage"* Ibid., 151.

---

53 *The most serious was* William J. Broad, "Oxygen Loss Causing Concern in Biosphere 2," *New York Times*, January 5, 1993.

53 *"The would-be Eden became a nightmare"* Ibid.

---

4. The Pinatubo Option

55 *His paper argued* Paul Crutzen, "Albedo Enhancement by Stratospheric Sulfur Injections: A Contribution to Resolving Policy Dilemma?" *Climatic Change* 77 (2006): 211.

56 *There was a natural analogue* Ibid.

57 *"Geoengineering is being discussed"* Editorial comment, "The Geoengineering Dilemma: To Speak or Not to Speak," Ibid., 247.

57 *the science section of the* William J. Broad, "How to Cool a Planet (Maybe)," *New York Times*, June 26, 2006, www.nytimes.com/2006/06/27/science/earth/27cool.html?_r=2&pagewanted=1&oref=slogin.

57 *"The source of the proposal"* Richard A. Kerr, "Pollute the Planet for Climate's Sake?," *Science* 314 (2006): 401–403.

58 *make the sky bluer or whiter* K. Caldeira and L. Wood, "Global and Arctic Climate Engineering: Numerical Model Studies" *Philosophcal Transactions of the Royal Society A* 366 (2008): 4039.

59 *A tenth of an ounce* David Keith, Presentation to National Academy of Sciences Geoengineering Workshop, Washington, D.C., June 14, 2009.

59 *the Royal Society noted* Royal Society, "Geoengineering the Climate: Science, Governance and Uncertainty" (2009): 32.

60 *"Geoengineering and Nuclear Fission"* Chris Mooney, "Can a Million Tons of Sulfur Dioxide Combat Climate Change?," *Wired*, June 23, 2008.

60 *wreaking possible ecological havoc* Bala Govindasamy and Ken Caldeira, "Geoengineering Earth's Radiation Balance to Mitigate CO_2-Induced Climate Change," *Geophysical Research Letters* 27 (2000): 2141.

60 *a relatively simplistic model* Ibid.

60 *"melting of Greenland"* Ibid.

61 *On Wednesday evening* Clive Oppenheimer, "Climatic, Environmental and Human Consequences of the Largest Known Historic Eruption: Tambora Volcano (Indonesia) 1815," *Progress in Physical Geography* 27 (2003): 234.

61 *Explosions were heard* Richard Stothers, "The Great Tambora Eruption in 1815 and Its Aftermath," *Science* 224 (1984): 1191.

61 *pillars of fire soared miles* Oppenheimer, "Climatic, Environmental and Human Consequences," 249.

61 *A 12-foot tsunami* Stothers, "The Great Tambora Eruption," 1191.

61 *endless stream of sulfur dioxide* Oppenheimer, "Climatic, Environmental and Human Consequences," 255.

61 *200 million tons* Ibid., 256.

62 *known by Westerners* Ibid., 244.

62 *the average temperature dropped* Ibid.

62 *Snow fell in June* Ibid.

62 *length of the growing season* Ibid., 245.

62 *"On the 10th of June"* Stothers, "The Great Tambora Eruption," 1196.

62 *Before there was Tambora* Alan Robock, "Volcanic Eruptions and Climate," *Reviews of Geophysics* 38 (2000): 191.

62 *famine in Rome and Egypt* J. P. Grattan and F. B. Pyatt, "Volcanic Eruptions Dry Fogs and the European Palaeoenvironmental Record: Localised Phenomena or Hemispheric Impacts?," *Global and Planetary Change* 21(1999): 173–179.

62 *the 1873 eruption of Lakagigar* Robock, "Volcanic Eruptions and Climate," 191.

62 *only a tenth as much rock* Oppenheimer, "Climatic, Environmental and Human Consequences," 1.

63 *requiring 180,000 flights per year* Alan Robock, Allison B. Marquardt, Ben Kravitz, and Georgiy Stenchikov, "The Benefits, Risks, and Costs of Stratospheric Geoengineering," *Geophysical Research Letters,* 36 (2009).

63 *"an annoying form of trash rain"* "Policy Implications of Greenhouse Warming: Mitigation, Adaptation, and the Science Base," Report of the Panel on Policy Implications of Greenhouse Warming, National Academy of Sciences, National Academy of Engineering, Institute of Medicine (Washington, D.C.: National Academy Press, 1992), 824.

67 *freshwater dumped into the oceans* K. E. Trenberth and A. Dai, "Effects of Mount Pinatubo Volcanic Eruption on the Hydrological Cycle as an Analog of Geoengineering," *Geophysics Research Letters* 34 (2007).

67 *A separate analysis* Alan Robock et al., "Regional Climate Responses to Geoengineering with Tropical and Arctic S02 Injections," *Journal of Geophysical Research* 113 (2008).

69 *The models are even worse* Edward Sarachik, Testimony to U.S. Senate Commerce Committee, May 8, 2008.

69 *drought over eastern Brazil* Phil Rasch et al., "An Overview of Geoengineering of Climate Using Stratospheric Sulfate Aerosols," *Philosophical Transactions of the Royal Society* 366 (2008): 4007.

69 *Perhaps most disturbingly* Lennart Bengtsson, "Geoengineering to Confine Climate Change: Is It at All Feasible?," *Climatic Change* 77 (2006): 229.

70 *"do not anticipate catastrophic changes"* Simone Tilmes et al., "Impact of Geoengineering Aerosols on the Troposphere and Stratosphere," *Journal of Geophysical Research* 114 (2009).

70 *A 2009 study estimated* Lina Mercato, "Impact of Changes in Diffuse Radiation on the Global Land Carbon Sink," *Nature* 458 (2009).

71 *The Pinatubo eruption* Lianhong Gu et al., "Response of a Deciduous Forest to the Mount Pinatubo Eruption: Enhanced Photosynthesis," *Science* 299 (2003).

71 *As Earth's forests were enhanced* Ibid., 2035.

71 *And yet ecologist Tony Janetos* Tony Janetos, presentation to the National Academy of Sciences geoengineering meeting, June 2009.

72 *dropped annual solar output by 14 percent* "Atmospheric Sunshade Could Reduce Solar Power Generation," Earth Science Research Library News, March 11, 2009, www.esrl.noaa.gov/news/2009/aerosol_implications_for_solar_power.html.

73 *Once upon a time* Kevin Trenberth, "Geoengineering: What, how, and for whom?," *Physics Today* (February 2009), 10.

74 *"Employing geoengineering schemes"* H. Damon Matthews and Ken Caldeira, "Transient Climate—Carbon Simulations of Planetary Geoengineering," *Proceedings of the National Academy of Sciences* 104 (2007): 9949.

74 *Matthews and Caldeira ran* Ibid.

74 *the "dystopic world" in the 1983 film* Joe Romm, "Exclusive: Caldeira Calls the Vision of Lomborg's Climate Consensus "a Dystopic World out of a Science Fiction Story," *Climate Progress*, September 5, 2009, http://climateprogress.org/2009/09/05/caldeira-delayer-lomborg-copenhagen-climate-consensus-geoengineering/.

---

75 *algae had turned the surface* Wendy Pyper, "Stocking Experiment Tests the Limit of the Lake," *Ecos*, April/June 2003, www.ecosmagazine.com/?act=view_file&file_id=EC115p4.pdf.

---

5. The Pursuit of Levers

77 *Swedish physicist Svante Arrhenius* Svante Arrhenius, "On the Influence of Carbonic Acid in the Air upon the Temperature of the Ground," *Philosophical Magazine* 41 (1896): 237–276.

77 *Respected meteorologist James Pollard Espy* James Fleming, "The Climate Engineers," *Wilson Quarterly* (Spring 2007): 51.

77 *a prescient U.S. government report* President's Science Advisory Committee, "Restoring the Quality of Our Environment," Executive Office of the President (Washington, D.C., 1965).

78 *flowing continents and an ever-changing* "History of Plate Tectonics," NASA, http://scign.jpl.nasa.gov/learn/plate2.htm.

79 *heat flows through the sky* James Fleming, "The Pathological History of Weather and Climate Modification: Three Cycles of Promise and Hype," *Historical Studies in the Physical and Biological Sciences*, 37, no. 1 (2006): 3–25.

79 *pines as thick as a man's leg* Ibid.

79 *former general Daniel Ruggles* Ibid., 7.

79 *Arrhenius's wife, Sofia* Wallace S. Broecker and Robert Kunzig, *Fixing Climate: What Past Climate Changes Reveal about the Current Threat—and How to Counter It* (New York: Hill & Wang, 2009), 69.

80 *Cutting the number* Ibid., 66.

80 *In his 1908 book* Ibid., 69.

80 *In 1901 he proposed* Nils Ekholm, "On the Variations of the Climate of the Geological and Historical Past and Their Causes," *Quarterly Journal of the Royal Meteorological Society* 27 (1901), 1–61.

80 *It seems possible that* Ibid.

81 *Arrhenius believed Earth* Fred Pearce, "With Speed and Violence: Why Scientists Fear Tipping Points in Climate Change," (2008), 5.

81 *Among other reasons* James Fleming, "Global Climate Change and Human Agency," in *Intimate Universality: Local and Global Themes in the History of Weather and Climate*,

edited by J. R. Fleming, V. Jankovic, and D. R. Coen (Sagamore Beach, Mass.: Science History Publications/USA, 2006), 236.

81 *When climatologist C. E. P. Brooks* Spencer Weart, "The Discovery of Rapid Climate Change," *Physics Today* 56 (2003), 30–36.

81 *"several thousand years to build"* Ibid.

82 *construction of the Hoover Dam* U.S. Bureau of Reclamation, "A Brief History of the Bureau of Reclamation," www.usbr.gov/history/BRIEFHist.pdf.

82 *In the 1890s Dokuchaev* Interview with Andre Lapenis, July 20, 2009.

82 *His book,* The Biosphere L. Margulis and D. Sagan, *What Is Life?* (Cambridge, Mass.: MIT Press, 1995).

83 *"Within days, a schematic map"* Mark Walker, *Science and Ideology: A Comparative History* (New York: Routledge, 2003), 142.

83 *The most prominent climatologist* Spencer Weart, "Climate Modification Schemes," *The Discovery of Global Warming* (American Institute of Physics, 2008), www.aip.org/history/climate/RainMake.htm.

84 *He used a simple climate model* Richard Kerr, "A Refuge for Life on Snowball Earth," *Science* 288 (2000), 1316.

84 *"We are merely on the threshold"* Nikolai Rusin and Liya Flit, *Man Versus Nature* (London: Central Books, 1966), 175.

84 *An experiment in 1960* David Keith, "Geoengineering the Climate: History and Prospect," *Annual Review of Energy Environment* 25 (2000): 250.

84 *warm northern Asia by more than* P. M. Borisov, "Can We Control the Arctic Climate?," *Bulletin of the Atomic Scientists* (March 1969): 43.

84 *President John F. Kennedy* Spencer Weart, "Climate Modification Schemes," *The Discovery of Global Warming* (American Institute of Physics, 2008), www.aip.org/history/climate/RainMake.htm.

85 *"The temperature of the air"* Nikolai Rusin and Liya Flit, *Man Versus Nature* (London: Central Books, 1966), 175.

85 *"If we want to improve our planet"* Ibid., 17.

85 *"A single pellet of dry ice"* *New York Times*, November 15, 1946, 24, as quoted in Fleming, "Pathological History," 3–25.

86 *"cloud seeding nuclei"* James Fleming, "Fixing the Weather and Climate: Military and Civilian Schemes for Cloud Seeding in Climate Engineering," in *The Technological Fix*, edited by Lisa Rosner (New York: Routledge, 2004).

86 *California Institute of Technology meteorologist* Kristine C. Harper, "Climate Control: United States Weather Modification in the Cold War and Beyond," *Endeavour* 32 (2008): 20–26.

86 *Strategic Air Command commander* *New York Times*, June 15, 1947, as quoted in Fleming, "Pathological History," 3–25.

86 *a "new race with the Reds"* "The Weather Weapon: New Race with the Reds," *Newsweek*, January 13, 1958.

86 *By 1960 annual federal support* David Keith, "Geoengineering the Climate: History and Prospect," *Annual Review of Energy Environment* 25 (2000): 245–284.

87 *A better solution* Morris Neiburger, "Weather Modification and Smog," *Science* 126 (1957): 637.

87 *Project STORMFURY* H. E. Willoughby, D. P. Jorgensen, R. A. Black, and S. L. Rosenthal, "Project STORMFURY: A Scientific Chronicle, 1962–1983," *Bulletin of the American Meteorological Society* 66 (1985): 505–514.

87 *By 1951 the young industry* Keith, "Geoengineering the Climate," 245–284.

87 *"Viet Cong safe and unsinged"* "South Viet Nam: Taking the Initiative," *Time*, April 9, 1965, www.time.com/time/magazine/article/0,9171,898584-2,00.html.

88 *Scientists were aghast* Fleming, "Fixing the Weather and Climate."

88 *one expert estimates* William Travis, "Geoengineering the Climate: Lessons from Purposeful Weather and Climate Modification," presentation, National Academies of Science Geoengineering Workshop, June 2009, http://americasclimatechoices.org/Geoengineering_Input/attachments/Travis_geoengineering_ACC.pdf.

88 *"predictable, detectable and verifiable results"* Michael Garstang et al., "Critical Issues in Weather Modification Research" (Washington, D.C.: National Academies Press, 2003).

89 *"Scientists had rejected old tales"* Spencer Weart, "The Discovery of Rapid Climate Change," *Physics Today* 56 (2003): 30.

89 *That conflicted with "the usual view"* David Ericson, 1956, as quoted in ibid.

89 *A twenty-seven-year-old Carl Sagan* Carl Sagan, "The Planet Venus," *Science* 133 (1961): 858.

89 *Twelve years later* Carl Sagan, "Planetary Engineering on Mars," *Icarus* 20 (1973): 513.

90 *in a government report on climate* President's Science Advisory Committee, "Restoring the Quality of Our Environment," Executive Office of the President (Washington, D.C., 1965).

90 *Budyko used a simple heat model* Mikhail Budyko, "The Effect of Solar Radiation Variations on the Climate of the Earth," *Tellus* 21 (1969).

91 *"destruction of polar ice"* Mikhail Budyko, *Climate and Life* (New York: Academic Press, 1974), 489.

91 *A year later, dismayed* Mikhail I. Budyko, *Climatic Changes* (1972; repr., Baltimore: Waverly Press, 1977).

91 *theorized that the refrigerant Freon* Seth Cagin and Philip Dray, *Between Earth and Sky* (New York: Random House, 1993), 66.

92 *"move the whole planet as a lever"* Ralph Cicerone, interview with author, July 21, 2009.

92 *Two years later he and colleagues* Ralph Cicerone et al., "Reduced Antarctic Ozone Depletions in a Model with Hydrocarbon Injections," *Science* 254 (1991): 1191.

92 *"That's the amplifier we need!"* Robert Charlson, interview with author, July 17, 2009.

92 *tiny amounts of the chemical* Robert Charlson, James Lovelock, Meinrat Andreae and Stephen Warren, "Oceanic Phytoplankton, Atmospheric Sulphur, Cloud Albedo, and Climate," *Nature* 326 (1987): 655.

93 *Four years later* Anthony Slingo, "Sensitivity of the Earth's Radiation Budget to Changes in Low Clouds," *Nature* 343 (1990): 49.

95 *Among scientists it was Teller* "Defense: Knowledge Is Power," *Time*, November 18, 1957, www.time.com/time/magazine/article/0,9171,868002,00.html.

95 *The program envisioned using nukes* "Executive Summary: Plowshare Program," U.S. Department of Energy, 2000, www.osti.gov/opennet/reports/plowshar.pdf.

95 *Nicaragua, Southeast Asia, and Egypt* M. C. MacCracken and J. R. Albritton, eds., 1992: *Atmospheric and Geophysical Sciences Program Report 1990-1991*, Lawrence Livermore National Laboratory report UCRL-51444-90-91 (Livermore, Calif.: 1991).

95 *"If your mountain is not in the right place"* Peter Goodchild, *Edward Teller, the Real Dr. Strangelove* (Cambridge, Mass.: Harvard University Press, 2004), 262c.

95 *Teller did defeat his activist rivals* Ibid., 318.

96 *during the Manhattan Project* Interview with Michael MacCracken, 2009.

96 *bombs that released less radioactivity* MacCracken and Albritton, eds., *Atmospheric and Geophysical Sciences Program Report 1990-1991*, 11.

96 *"The study reinforced the need"* Ibid.

97 *clashed with Linus Pauling* Goodchild, *Edward Teller*, xxiii.

97 *"I was the only victim"* Edward Teller, in *Wall Street Journal,* 1979, as quoted in Stanley Bloomberg and Lewis Panos, *Edward Teller* (New York: Macmillan, 1990).

97 *the program would cost merely* E. Teller, L. Wood, and R. Hyde, "Global Warming and Ice Ages: I. Prospects for Physics-based Modulation of Global Change," August 15, 1997, Lawrence Livermore National Laboratory #UCRL-JC-128715.

97 *He followed it up with an op-ed* Edward Teller, "The Planet Needs a Sunscreen," *Wall Street Journal*, October 17, 1997.

98 *In 2002 Teller expanded the idea* E. Teller, R. Hyde, and L. Wood, "Active Climate Stabilization: Practical Physics-Based Approaches to Prevention of Climate Change," National Academy of Engineering Symposium, Washington, D.C., April 23–24, 2002, 1.

99 *"until the next Ice Age"* Ibid., 6.

---

101 *Hearing this advice* Tina Butler, "Overstaying Their Welcome: Cane Toads in Australia," Mongabay.com, April 17, 2005.

102 *In 2005 an Australian lawmaker* "Cane Toad Clubbing Sparks Controversy," *ABC News online*, April 11, 2005, www.abc.net.au/news/newsitems/200504/s1342444.htm.

---

6. The Sucking-1-Ton Challenge

103 *Humanity emits 30 billion tons* Energy Information Administration Report 0484, May 27, 2009, www.eia.doe.gov/oiaf/ieo/emissions.html.

103 *Until 2001* Kera Abraham, "Moss Landing Inventor Makes Cement a Tool against Global Warming," *Monterey County Weekly*, September 11, 2008, www.montereycounty weekly.com/archives/2008/2008-Sep-11/moss-landing-inventor-makes-cement-a-tool-against-global-warming/1/.

104 *explosives during World War II* Ibid.

104 *most carbon-intensive materials* Lisa J. Hanle and K. Jayaraman, "CO_2 Emissions Profile of the U.S. Cement Industry," Report for the 13th International Emission Inventory Conference, Clearwater, Florida, June 8, 2004, www.epa.gov/ttnchie1/conference/ei13/ghg/hanle.pdf, 10.

105 *more than 2,100 coal plants* Eli Kintisch, "Power Generation: Making Dirty Coal Plants Cleaner," *Science* 317 (2007): 184–186.

105 *41 percent of total world CO_2 emissions* "The Future of Coal," An Interdisciplinary MIT Study, Summary Report, 2007, http://web.mit.edu/coal/The_Future_of_Coal_Summary_Report.pdf.

105 *emit a whopping 10 percent* Kurt House, "Electrochemical Acceleration of Chemical Weathering," presentation to the Kavli Institute for Theoretical Physics Conference: Frontiers of Climate Science, May 9, 2008, http://online.kitp.ucsb .edu/online/climate_c08/house/.

105 *Coal is cheap and abundant* "The Future of Coal," ix.

105 *half the electricity generated* Ibid.

106 *"If we don't solve the climate problem"* Kintisch, "Power Generation," 184–186.

106 *could double the average electric bill* Dale Simbeck, interview with the author, December 6, 2009.

107 *experts calculated* Malte Meinshausen, "Greenhouse-Gas Emission Targets for Limiting Global Warming to 2°C," *Nature* 458 (2009).

107 *If it wasn't for the greenhouse effect* H. Le Treut et al., "2007: Historical Overview of Climate Change," in *Climate Change 2007: The Physical Science Basis. Contribution of Working Group I to the Fourth Assessment Report of the Intergovernmental Panel on Climate Change*, ed. S. Solomon, D. Qin, M. Manning, Z. Chen, M. Marquis, K. B. Averyt, M. Tignor, and H. L. Miller (Cambridge, U.K.: Cambridge University Press, 2007), 97.

108 *Carbon dioxide makes up only* "The Future of Coal," 20.

109 *transformation of the world's coal power plants* Armond Cohen, Mike Fowler, and Kurt Waltzer, "NowGen: Getting Real about Coal Carbon Capture and Sequestration," *Electricity Journal* 22 (2009): 27.

109 *demonstration projects in the United States* Ibid., 32.

109 *roughly $3,730 per kilowatt* "Duke Energy Revises Cost Estimate for Edwardport," *Inside Indiana Business*, May 2, 2008.

110 *one to three centuries* Union of Concerned Scientists, "How Coal Works," www.ucsusa .org/clean_energy/coalvswind/brief_coal.html (accessed October 7, 2009).

110 *equal in generating capacity* MIT Energy Initiative Symposium, "Retrofitting of Coal-Fired Plants for CO_2 Emissions Reductions" March 23, 2009, http://web.mit.edu/mitei/ docs/reports/meeting-report.pdf.

110 *most of this growth from China and India* Ibid.,15.

110 *pure oxygen in the boiler* Scottish Center for Carbon Storage, "Oxy-Fuel Combustion Capture," www.geos.ed.ac.uk/sccs/capture/oxyfuel.html (accessed October 7, 2009).

110 *method requires loads of energy* "The Future of Coal," 32.

111 *The so-called parasitic load* Kintisch, "Power Generation," 184–186.

112 *California expert Dale Simbeck* Dale Simbeck and Waranya Roekpooritat, "Near-Term Technologies for Retrofit CO_2 Capture and Storage of Existing Coal-Fired Power Plants in the United States," paper for the MIT Retrofit Symposium, May 2009, http://web.mit .edu/mitei/docs/reports/simbeck-near-term.pdf.

112 *Scientists working at Notre Dame* National Energy Technology Laboratory, "Ionic Liquids: Breakthrough Absorbtion Technology for Post-Combustion CO_2 Capture," project summary, August 2007, www.netl.doe.gov/publications/factsheets/project/ Proj471.pdf.

113 *each sucking station pulling in* Peter Fairley, "Algerian Carbon Capture Success," *Technology Review* blog, December 19, 2008, www.technologyreview.com/blog/ energy/22468/.

113 *To get all that carbon* Nicola Jones, "Climate Crunch: Sucking It Up," *Nature* 458 (2009): 1095.

114 *unfettered by carbon regulations* Roger A. Pielke Jr., "An Idealized Assessment of the Economics of Air Capture of Carbon Dioxide in Mitigation Policy," *Environmental Science and Policy* 12 (2009): 216–225.

114 *phalanx of air-capture devices* Eli Kintisch, "Wanted: Suckers," *Science Insider* blog, February 11, 2009, http://blogs.sciencemag.org/scienceinsider/2009/02/wanted-suckers .html.

115 *Swiss scientists are trying* Jones, "Climate Crunch," 1095.
116 *an American company called Rentech* Eli Kintisch, "Energy: The Greening of Synfuels," *Science* 320 (2008): 306–308.
117 *an alternative to gasoline* Ibid.
118 *20 million barrels a day* "The Future of Coal," ix.
118 *target would be porous rocks* Ibid., 44.
118 *more than 9* billion *tons of* CO_2 Ibid., 43.
119 *"risks appear small"* Ibid., 50.
119 *fluid underground might cause earthquakes* Peter Folger, "Underground Carbon Dioxide Sequestration: Frequently Asked Questions," Congressional Research Service, January 21, 2009, http://web.mit.edu/lugao/MacData/afs/sipb.mit.edu/contrib/wikileaks-crs/wikileaks-crs-reports/RL34218.pdf, 8.
119 *known as the Schwarze Pumpe* Roger Harrabin, "Germany Leads 'Clean Coal' Pilot," *BBC News*, September 3, 2008, http://news.bbc.co.uk/2/hi/science/nature/7584151.stm.
119 *"questions about the safety"* Terry Slavin and Alok Jha, "Not under Our Backyard, Say Germans, in Blow to CO_2 Plans," Guardian.co.uk, July 29, 2009.
120 *citing "business considerations."* Ben Sutherly, "Greenville Carbon Dioxide Injection Plan Abandoned," *Dayton Daily News*, August 20, 2009.
120 *Constantz attacked geologic sequestration* Brent Constantz, Testimony before the U.S. Senate, Appropriations Committee, Subcomittee on Energy and Water Development, May 6, 2009.
120 *The global cement industry emits* World Resources Institute, "CO_2 Emissions by Source," http://earthtrends.wri.org/text/climate-atmosphere/variable-465.html (accessed October 7, 2009).
121 *In the first two years of its existence,* Calera Corporation, "Sequestering CO_2 in the Build Environment," presentation to the MIT Energy Initiative Symposium, March 23, 2009, http://web.mit.edu/mitei/docs/reports/calera-sequestering.pdf, slide 4.
121 *Khosla has poured more* Brent Constantz, August 3, 2009, interview with author.
122 *"sign a nondisclosure agreement."* Ken Caldeira, post to Climate Intervention Google group, March 24, 2009, http://groups.google.com/group/climateintervention/msg/a1d25a87508a290c?pli=1.

---

127 *"Tires, which were an aesthetic"* David Fleshler, "States Attempt to Clean Up after a Failed Artificial Reef," *South Florida Sun Sentinel*, July 18, 2003.
127 *Fish never inhabited* Ibid.
127 *The effort to clean up* Jim Loney, "Florida Raises Ill-Fated Artificial Reefs," Reuters, July 9, 2007.

---

7. Credit Is Due

129 *"Voyage of Recovery"* "Planktos Launches Galápagos 'Voyage of Recovery' Green Climate Initiative." Planktos press release, March 12, 2007.
131 *In a 1993 experiment* J. H. Martin et al., "Testing the Iron Hypothesis in Ecosystems of the Equatorial Pacific Ocean," *Nature* 371 (1994): 123–129.
132 *Some scientists believe* Victor Smetacek, e-mail to author, September 15, 2008; Margaret Leinen, e-mail to author, September 15, 2008.

133 *worth roughly $25 per ton* Joseph Romm, "Do the 2 Billion Offsets Allowed in Waxman Markey Gut . . ." Climate Progress blog, http://climateprogress.org/2009/05/27/domestic-international-offsets-waxman-markey/.

135 *algae to boost fish stocks* Aaron Strong et al., "Ocean Fertilization Science, Policy, and Commerce," *Oceanography* 22 (2009): 236–261.

135 *"30 percent of the CO_2 produced"* Mike Markels, "Farming the Oceans: An Update," *Regulation* 21 (2): 9–10.

135 *Four scientists wrote in* Nature Aaron Strong et al., "Ocean Fertilization: Time to Move On," *Nature* 461 (2009): 347.

135 *MIT ecologist Sallie Chisholm* Sallie Chisholm et al., "Discrediting Ocean Fertilization," *Science* 294 (2001): 309.

137 *Planktos's main investor* *R. v. Skalbania*, 3 Supreme Court of Canada Reports 995, November 6, 1997.

137 *"We have shipping agents"* "Seeding the Ocean with Iron Runs into Opposition," *Ottawa Citizen*, June 19, 2007.

138 *selling it for $750 million* Robin Lancaster, "Upon a Painted Ocean," *Point Carbon* (April 2008).

140 *"French maid feather duster"* Melodie Grubb, Planktos *Weatherbird* blog, February 19, 2007, http://planktosweatherbird.blogspot.com/ (accessed October 7, 2009).

141 *"sovereignty of our fisherfolk"* Stephen Leahy, "Ecuador: Doubts Surround Carbon Absorption Project Near Galápagos," IPSNews.net http://ipsnews.net/news.asp?idnews=38533 (accessed October 9, 2009).

141 *"I Am Not the Enemy"* Russ George, "I Am Not the Enemy," *Ottawa Citizen*, June 19, 2007.

143 *"This is YOUR opportunity"* *Global Warming, Inc.*, "GroundFloor Stocks," August 2007.

144 monitored water around oil rigs Nelson Skalbania, interview with author, December 6, 2009, http://csc.lexum.umontreal.ca/en/1997/1997scr3-995/1997scr3-995.html.

144 *In 1995, at age twenty-nine* Lorna Fernandes, "It's No Small World for ITN," *Silicon Valley/San Jose Business Journal*, October 27, 1997.

145 *"barely squeaked by"* Ibid.

146 *they would set up rules* Eli Kintisch, "Rules for Ocean Fertilization Could Repel Companies," *Science*, November 7, 2008.

148 *"a fox in the henhouse"* Eli Kintisch, "March Geoengineering Confab Draws Praise, Criticism," ScienceInsider blog, November 6, 2009, http://blogs.sciencemag.org/scienceinsider/2009/11/march-geoengine.html.

---

149 *Soviet engineers embarked* Richard Stone, "Coming to Grips with the Aral's Grim Legacy," *Science* 284 (1999): 30.

149 Science *called the* Christopher Pala, "Once a Terminal Case, the North Aral Sea Shows New Signs of Life," *Science* 312 (2006): 183.

---

8. Victor's Garden

151 *"merry-go-round ecosystems"* Victor Smetacek, "The 'Merry-go-Round Ecosystems' of Antarctica: A Study in Stability," opening plenary lecture, 7th Scientific

Committee on Antarctic Research, International Biology Symposium, Christchurch, New Zealand (1998).

151 *They continually recycle nutrients* John Martin et al., "Iron in Antarctic Waters," *Nature* 345 (1990): 156–158.

153 *perhaps the finest research vessel Polarstern* technical data, Alfred Wegener Institute, www.awi.de/en/infrastructure/ships/polarstern/technical_data/.

154 *"My father had been a sea-farer"* Victor Smetacek, "Personal Background," Karen Smetacek home page, www.smetacek.de/victor.htm.

154 *"Can you expect"* Bill Davidson, "Bread from the Sea," *Reader's Digest*, June 1954.

154 *"Territorial males hate to be"* "A Voyage of Discovery: Victor Smetacek," *Nature* 421 (2003): 897, http://natureeducation.com/nature/journal/v421/n6926/full/421897a.html.

154 *some left-handed people* Victor Smetacek, "Mirror-Script and Left-Handedness," *Nature* 355 (1992): 118–119.

155 *1 billion tons of* CO_2 Eli Kintisch, "Debate: Do Gobbled Algae Mean Carbon Fix Sunk?," ScienceInsider blog, March 31, 2009, http://blogs.sciencemag.org/scienceinsider/2009/03/debate-do-gobbl-1.html.

156 *"We shall have to watch our weights"* Victor Smetacek, LohaFex circular, December 2009, www.biodiv-network.de/upload/presse/deutsch/smetacek_naqvi.pdf.

157 *"controversial climate change experiment"* Bobby Jordan, "Rogue Ship Sails into Storm over Experiment," *Times of South Africa*, January 11, 2009.

158 *"rules to prevent rogue geoengineers"* "LOHAFEX Update: Geo-engineering Ship Plows on as Environment Ministry Calls for a Halt," ETC Group, January 13, 2009, www.etcgroup.org/en/node/712.

159 *"So adding a few hundred gigatons"* Victor Smetacek and Wahid Naqvi, www.biodiv-network.de/upload/presse/deutsch/smetacek_naqvi.pdf.

160 *in the direction of Tierra del Fuego* Stefan Schwarz, speech to *Polarstern*, March 15, 2009.

163 *vital signs such as temperature* Wajih Naqvi and Victor Smetacek, "ANT XXV/3, Weekly Report No. 8," LohaFex, March 19, 2009.

163 *a slowly rotating whirlpool* LohaFex press release, March 23, 2009.

166 *"it has dampened hopes on the potential"* Ibid.

166 *"But our results show"* Richard Black, "Setback for Climate Technical Fix," BBC News, March 23, 2009, http://news.bbc.co.uk/2/hi/science/nature/7959570.stm.

---

169 *Les Kauffman of* Nancy Chege, "Lake Victoria: Sick Giant," *People and the Planet*, www.cichlid-forum.com/articles/lake_victoria_sick.php.

---

9. The Sky and Its Reengineer

172 *roughly 2 million gallons of water an hour* Stephen Salter, "Spray Turbines to Increase Rain by Enhanced Evaporation from the Sea," 10th Congress of International Maritime Association of the Mediterranean, May 2002.

172 *trillion trillion droplets* John Latham, interview with author, May 2009.

173 *three hundred tons each* Stephen Salter, "Seagoing Hardware for the Cloud Albedo Method of Reversing Global Warming," *Philosophical Transactions of the Royal Society A* 366 (2008): 4001.

173 *The British Ministry of Aviation* Andrew McPhee, www.unrealaircraft.com/classics/sr_53_177.php.

174 *He stumbled on* Anthony Slingo, "Sensitivity of the Earth's Radiation budget to Changes in Low Clouds," *Nature* 343 (1990): 49.

174 *"stratocumulus" clouds* Alan Gadian, "Reflecting the Sunlight from Stratus Clouds," presentation to nonprofit groups, Washington, D.C., May 2009.

175 *"Control of Global Warming?"* John Latham, "Control of Global Warming?," *Nature* 347 (1990): 339.

175 *A computer modeling study* Keith Bower et al., "Computational Assessment of the Proposed Technique for Global Warming Mitigation via Albedo Enhancement of Marine Stratocumulus Clouds," *Atmospheric Research* 82 (2006): 328.

176 *"Are you so confident"* Stephen Salter, "Comment on the Letter from DEFRA to Salter of 15 November 2005," www.see.ed.ac.uk/~shs/Climate%20change/Albedo%20control/Reply%20to%20DEFRA%202.pdf.

180 *a million million water drops* John Latham, interview with author, March 2009.

180 *coagulate and form larger ones* Johann Feichter, e-mail message to author, April 6, 2009.

187 *save "New York from the next Katrina"* Stephen Salter, "Royal Society Call for Submissions on Geoengineering: Submission on Wave-Powered Sinks for Hurricane Suppression, Carbon Removal, and Enhanced Phytoplankton Growth," October 2008.

187 *the British Royal Society published a report* "Geoengineering the Climate," British Royal Society, September 2009.

---

189 *Spindly with purple flowers* Jim Robbins, "A Weed, a Fly, a Mouse and a Chain of Unintended Consequences," *New York Times,* April 4, 2006.

189 *research published in 2006* Dean E. Pearson and Ragan M. Callaway, "Biological Control Agents Elevate Hantavirus by Subsidizing Deer Mouse Populations," *Ecology Letters* 9 (2006): 443.

---

10. The Right Side of the Issue

193 *"taints" of modern technological life* Alvin Weinberg, *Reflections on Big Science* (Cambridge, Mass.: MIT Press, 1967), 35.

194 *In a 2009 column* George Will, "Cooling down the Cassandras," *Washington Post*, October 1, 2009.

194 *"Seizing upon either the low end"* Jonathan Chait, "The Two Arguments against Reducing Emissions," The Plank, *New Republic* Web site, October 2, 2009.

196 *the greenhouse gases we've already emitted* Stephen Dubner and Steven Levitt, *SuperFreakonomics: Global Cooling, Patriotic Prostitutes, and Why Suicide Bombers Should Buy Life Insurance* (New York: William Morrow, 2009), 188.

196 *a "costly, complicated" solution* Stephen Dubner and Steven Levitt, "More Than One Way to Cool Earth," *USA Today*, October 27, 2009.

196 *"fiendishly simple plan"* Dubner and Levitt, *SuperFreakonomics*, 195.

196 *He called it "delightful."* Bret Stephens, "Freaked Out over SuperFreakonomics," *Wall Street Journal*, October 27, 2009.

197 *"Conservative environmentalism" believes* Robert Locke, "The Right Conservative Position on the Environment," FrontPageMagazine.com, November 2, 2001.

197 *After George W. Bush was elected* Ehsan Khan, communication with author, May 2009.

198 *"There is a significant risk"* "Response Options to Limit Rapid or Severe Climate Change: Assessment of Research Needs," draft white paper, U.S. Department of Energy, Washington, D.C., October 2001.

198 *failing to establish a sufficient cadre* Rowan Scarborough, "Lack of Translators Hurts U.S. War on Terror," *Washington Times*, August 31, 2009.

199 *does not "necessarily" warm Earth* Dubner and Levitt, *SuperFreakonomics* 183.

199 *"So hopelessly wrong"* William Connolley, "SuperFreakonomics: Global Cooling (and Some Other Stuff)?," STOAT blog, October 13, 2009, http://scienceblogs.com/stoat/2009/10/superfreakonomics_global_cooli.php.

200 *On a radio program* Steven Levitt, interview by Scott Simon, *Weekend Edition Saturday,* National Public Radio, October 17, 2009.

200 *in his analysis he argued* Alan Carlin, "Proposed NCEE Comments on Draft Technical Support Document," EPA NCEE, http://bit.ly/QB06u.

200 *"it would actually work."* Zachary Roth, "Climate Skeptic: 'I Was Hoping People at EPA Would Pay Attention to My Work,'" *Talking Points Memo*, July 1, 2009.

202 *"Geoengineering would provide more time"* Ronald Bailey, "An Emergency Cooling System for the Planet," Reason.com, June 10, 2008, http://reason.com/archives/2008/06/10/an-emergency-cooling-system-fo.

202 *"With regulations and rations"* Neil Reynolds, "Climate Cures: Technology, Not Ideology," *Toronto Globe and Mail*, June 13, 2008.

203 *Teller estimated that the Pinatubo Option* Edward Teller, "Sunscreen for Planet Earth," *Wall Street Journal*, October 15, 1997.

203 *puts the cost of deploying* Dubner and Levitt, *SuperFreakonomics*, 196.

203 *The 2009 report on geoengineering* J. Eric Bickel and Lee Lane, "An Analysis of Climate Engineering as a Response to Climate Change," Copenhagen Consensus Center report, August 7, 2009.

204 *"makes it clear that"* Copenhagen Consensus Center press release, September 4, 2009.

204 *In a companion paper, however* Roger A. Pielke Jr., "A Perspective Paper on Climate Engineering, Including an Analysis of Carbon Capture as Responses to Climate Change," Copenhagen Consensus Center, August 7, 2009.

205 *"How can you vote on which solutions"* Alvia Gaskill, message to Google Groups geoengineering message board, August 8, 2009.

207 *"giving serious consideration to geoengineering"* John Holdren, Office of Science and Technology Policy, personal e-mail, www.judicialwatch.org/files/documents/2009/OSTP_climate_response1_5_2009.pdf.

207 *"My reason for this is a political ploy"* Richard Courtney, comment on forum.junkscience.com (comment posted August 12, 2009).

207 *In April 2009* Michael Totty, e-mails to Ken Caldeira, April 27, 2009, and April 29, 2009.

208 *"It's Time to Cool the Planet"* Jamais Cascio, "It's Time to Cool the Planet," *Wall Street Journal*, June 15, 2009.

208 *"Delay is the Carbon Lobby's strategy"* Alex Steffen, "Geoengineering and the New Climate Denialism," *Worldchanging*, April 29, 2009, www.worldchanging.com/archives/009784.html.

---

209 *nongovernmental organizations* Maina Waruru, "Saviour tree turns scourge in Kenya," Science and Development Network News (SciDev.Net), November 24, 2009.

209 *Its leaves fertilize* Caroline Irby, "Devil of a Problem: the Tree that's Eating Africa," *The Independent*, London, August 27, 2004.

210 *"If there's a use"* Ibid.

---

11. A Political Climate

212 *"charitable, artistic, educational"* Colouste Gulbenkian Foundation, www.gulbenkian.pt/media/files/fundacao/historia_e_missao/PDF/STATUTES.pdf.

212 *"in a few decades the option of geoengineering"* David Victor et al., "The Geoengineering Option: A Last Resort against Global Warming?," *Foreign Policy* (March/April 2009).

218 *local environmentalists and fishermen* Mark Anthony de Figueiredo, "The Hawaii Carbon Dioxide Ocean Sequestration Field Experiment: A Case Study in Public Perceptions and Institutional Effectiveness" (graduate thesis, MIT, 2003).

218 *Subsequent experiments on a smaller scale* Peter Brewer, interview with author, July 7, 2009.

221 *"Chinese naval vessel armed"* Thomas C. Schelling, "The Economic Diplomacy of Geoengineering," *Climatic Change* 33 (1996): 303.

221 *"short-lived fluorocarbon greenhouse gases"* David S. Battisti et al., "Climate Engineering Responses to Climate Emergencies," NOVIM (June 12, 2009), 28, www.docstoc.com/docs/document-preview.aspx?doc_id=9676718.

222 *To illustrate the uncharted waters* Scott Barrett, "The Incredible Economics of Geoengineering," *Environmental and Resource Economics* 39 (2008): 53.

222 *"countries are pretty much free"* Ibid.

222 *The 1959 Antarctic Treaty* "The Antarctic Treaty," December 1, 1959, full text online, www.nsf.gov/od/opp/antarct/anttrty.jsp.

223 *tapping mineral resources there* Daniel Bodansky, "May We Engineer the Climate?," *Climatic Change* 33 (1996): 309–321.

224 *By the same token* Martin Bunzl, "Researching Geoengineering: Shouldn't or Couldn't?" Paper submitted to National Academy of Sciences, June 10, 2009.

---

229 *In the 1950s* Press release, Commonwealth Scientific and Industrial Research Organization, www.ento.csiro.au/publicity/pressrel/2001/13jun01.html.

229 *"I absolutely shun"* Ibid.

---

12. Geoengineering and Earth

231 *even half its warming potential* David Archer and Victor Brovkin, "The Millennial Atmospheric Lifetime of Anthropogenic CO_2," *Climatic Change* 90 (2008): 283–297.

232 *"splits early into two contrasting schools"* Johann Hari, "Move Over, Thoreau," *Slate*, January 12, 2009.

233 *"Greens are no longer strictly"* Steward Brand, *Whole Earth Discipline: An Ecopragmatist Manifesto* (New York: Viking, 2009).

233 *asserted in* A Sand County Almanac Aldo Leopold, *A Sand County Almanac* (Oxford, U.K.: Oxford University Press, 1949), 226.

233 *"A wilderness is where the flow"* David Brower, *Let the Mountains Talk, Let the Rivers Run* (New York: HarperCollins, 1995), 71.

233 *the "junkie logic"* Bill McKibben, quoted in in Computing for Sustainability blog, May 1, 2009, http://computingforsustainability.wordpress.com/2009/05/01/bill-mckibbens-350ppm-at-46-south/.

233 *"truly and viscerally think of ourselves"* Bill McKibben, *The End of Nature:10th Anniversary Edition* (New York: Random House, 1999), 146.

234 *"We are as gods,"* *Whole Earth Catalog (*Menlo Park, Calif.: Portola Institute, 1968), 1.

234 *"Whether it's called managing the Commons"* Brand, *Whole Earth Discipline, 275.*

234 "integrity, stability, and beauty" Aldo Leopold, "The Land Ethic," in *American Earth: Environmental Writing since Thoreau*, edited by Bill McKibben (New York: Penguin, 2008), 292.

234 *"We live in the oddest moment"* McKibben, *End of Nature,* xv.

235 *Yearly global greenhouse gas emissions* *Climate Change 2007*, the Intergovernmental Panel on Climate Change, Working Group 3.

235 *"forever in the deity business."* McKibben, *End of Nature, 166.*

235 *"we have retarded some part of man's"* Rachel Carson, address to Theta Sigma Phi, 1954.

235 *singing the praises of living "humbly"* McKibben, *End of Nature*, 64.

235 *"part of the food chain"* Ibid., 146.

236 *"Are we really suggesting"* Bill McKibben, quoted in Computing for Sustainability blog, May 1, 2009, http://computingforsustainability.wordpress.com/2009/05/01/bill-mckibbens-350ppm-at-46-south/.

236 *"Rather than applied engineering"* Bill Becker, "So Much for Geoengineering," Climate Progress blog, http://climateprogress.org/2009/02/12/geoengineering-bad-idea-iron-fertilization/.

236 *Visionary and futurist R. Buckminster Fuller* R. Buckminster Fuller, "Spaceship Earth," in *Operating Manual for Spaceship Earth* (Carbondale: Southern Illinois University Press, 1969), 49–56.

237 *"a world-sized problem"* Stewart Brand, 2009 interview with *Edge* magazine, August 20, 2009, www.edge.org/3rd_culture/brand09/brand09_index.html.

237 *"ideologies have to shift"* Brand, *Whole Earth Discipline*, 1.

237 *"Accustomed to saving natural systems"* Ibid., 20.

237 *"geoengineering should be 'safe, legal, and rare,'"* Ibid., 297.

237 *"Beavers do it"* Stewart Brand, "We Are as Gods and Have to Get Good at It," Edge .org, August 20, 2009, www.edge.org/3rd_culture/brand09/brand09_index.html#video.

238 *his 1979 classic,* Gaia James Lovelock, *Gaia: A New Look at Life on Earth* (Oxford, U.K.: Oxford University Press, 1979).

238 *"part of the natural scene"* Ibid., 127.

238 *"no conceivable hazard"* James Lovelock et al., "Halogenated Hydrocarbons in and over the Atlantic," *Nature* 241 (1973): 194–196.

239 *"one of my greatest blunders."* James Lovelock, *The Ages of Gaia: A Biography of Our Living Earth* (New York: W. W. Norton, 1988), 135.

239 *"When the urban industrial man"* Lovelock, *Gaia: A New Look*, 117.

239 *Lovelock said humans were* Ibid., 148.

239 *"doomed to failure"* Ibid., 145.

239 *"It is up to us to act"* Lovelock, *The Ages of Gaia*, 239.

239 *measures such as the Pinatubo Option* James Lovelock, *The Revenge of Gaia: Why the Earth Is Fighting Back and How We Can Still Save Humanity* (London: Allen Lane, 2006).

239 *"heart and mind of the Earth"* James Lovelock, "A Geophysiologist's Thoughts on Geoengineering," *Philosophical Transactions of the Royal Society A*, 366 (2008): 3883–3890.

239 *"Whatever our mistakes"* James Lovelock, *The Vanishing Face of Gaia* (New York: Basic Books, 2009).

240 *"Geoengineering is like trying"* James Lovelock, "Such Drastic Climate Therapy Could Make Things Worse," *Guardian*, September 20, 2009, www.guardian.co.uk/commentisfree/cif-green/2009/sep/20/geoengineering-royal-society-earth.

242 *Opponents of geoengineering* Alan Carlin, "Risky Gamble," *Environmental Forum* (September/October 2007), 42.

Index